CW00766951

Panzergrenadier Divisions

Panzergrenadier Divisions of the Waffen-SS

Rolf Michaelis

Schiffer Military History
Atglen, PA

Book translation by David Johnston

Book Design by Ian Robertson.

Library of Congress Control Number: 2010931601

Printed in China.
ISBN: 978-0-7643-3660-7

This book was originally published in German under the title
Die Panzergrenadier-Divisionen der Waffen-SS by Michaelis-Verlag

We are interested in hearing from authors with book ideas on related topics.

Published by Schiffer Publishing Ltd.
4880 Lower Valley Road
Atglen, PA 19310
Phone: (610) 593-1777
FAX: (610) 593-2002
E-mail: Info@schifferbooks.com.
Visit our web site at: www.schifferbooks.com
Please write for a free catalog.
This book may be purchased from the publisher.
Please include $5.00 postage.
Try your bookstore first.

In Europe, Schiffer books are distributed by:
Bushwood Books
6 Marksbury Avenue
Kew Gardens
Surrey TW9 4JF, England
Phone: 44 (0) 20 8392-8585
FAX: 44 (0) 20 8392-9876
E-mail: Info@bushwoodbooks.co.uk.
Visit our website at: www.bushwoodbooks.co.uk

Yellowjackets! The 361st Fighter Group in World War II - P-51 Mustangs over Germany *Paul B. Cora*. A narrative history of a combat unit attached to the U.S. 8th Air Force in the ETO from December 1943 through the end of the war in Europe. More than a narrow unit history, *Yellowjackets*! places the experiences of the Group's personnel in perspective with the larger context of the war, and also presents the human side, speaking, at times, from the intensely personal standpoint of the pilots and personnel.
Size: 8.5"x11" • over 130 b/w & color photographs, color aircraft profiles • 152 pp. • ISBN: 0-7643-1466-1 • hard cover • $39.95

The 370th Fighter Group in World War II - in Action over Europe with the P-38 and P-51 *Jay Jones*. The history of the 370th Fighter Group in World War II, conveyed in the words and photos of the veterans. The 370th was organized in 1943 to fly the P-47 Thunderbolt. When the group arrived in England in early 1944, they were assigned to the 9th Air Force and converted to fly P-38 Lightnings. They were involved in every major Allied offensive from D-Day onward. The group supported Operation Cobra, flew missions over the Falaise Gap, and flew cover over Operation Market Garden. On missions after the weather cleared, the men could actually watch from the base as their planes made dive-bombing runs on German armored columns. The group converted to P-51 Mustangs in March 1945, just in time for cover missions during Operation Varsity, the Rhine River crossing. This is one of the most thorough and comprehensive group histories written and is a must for veterans, their families, and enthusiasts.
Size: 8.5"x11" • over 700 black and white and color photos, color aircraft profiles • 448 pp. • ISBN: 0-7643-1779-2 • hard cover • $59.95

We are always interested in working with authors or collectors on wonderful book ideas. Contact us at **www.schifferbooks.com**, and page down to "For Authors – Submit a Book Proposal." Or call or write to us about your proposal.

Talk with your local bookseller. Their knowledge is one of the best tools in finding the correct book, and we all need to support them.

** Peter Schiffer and Team **
Designed & composed in America.
Printed in China.

Contents

Foreword

The experiences of the First World War had shown that a war could only be won through movement. In the Second World War it was the panzer divisions and the motorized infantry divisions, later reorganized and renamed panzer grenadier divisions, that were the instruments of this war of movement.

The panzer grenadier division was a mixture of panzer division and infantry division. In addition to a panzer battalion, the panzer grenadier division had a variety of units that were motorized, trucked, or armored. In the face of ever more serious fuel shortages, the panzer grenadier divisions were often more employable than panzer divisions because of their lower fuel consumption, and of course because of their equipment they possessed greater firepower than a pure infantry division.

The goal of this book is to tell the stories of the six Waffen-SS panzer grenadier divisions, describing their formation, service history, and personnel composition. A selection of personal accounts, plus photographs, documents, and division emblems complete the work.

This book describes the history of these units, but it does not show the misery that reigned in Europe during the Second World War. The reader should not forget this.

Berlin, July 2008
Rolf Michaelis

4th SS Panzer-Grenadier Division

(4. SS-Polizei-Panzergrenadier-Division)

The Police Division
(Polizei-Division)

On 18 September 1939 Hitler ordered the Chief of the Order Police (*Ordnungspolizei*) to form a horse-drawn infantry division from members of the police. The idea of using policemen in a military role was not new. In 1938-39 elements of the police, organized in battalions and groups, had taken part in the actions in Austria and Czechoslovakia.

Formation of the unit began at Training Camp "Wandern," near Frankfurt/Oder, under the command of *Generalleutnant der Polizei* Pfeffer-Wildenbruch. Designated the Police Division (*Polizei-Division*), the unit was to be organized as follows:

Division Headquarters
Police Rifle Regiment 1 (I – III Battalion)
Police Rifle Regiment 2 (I – III Battalion)
Police Rifle Regiment 3 (I – III Battalion)
Police Bicycle Troop (reconnaissance unit)
Police Anti-Tank Battalion
Police Pioneer Battalion
Police Supply Services
Police Medical Battalion
Police Veterinary Battalion

The division, whose "soldiers" remained members of the *Ordnungspolizei*, and which was only formed within the framework of the army, was also assigned:

Artillery Regiment 300 (Army)
    3 battalions (105-mm light field howitzers)
    1 battalion (150-mm heavy field howitzers)
Signals Battalion 300 (Army)
    1st Company (field telephone)
    2nd Company (radio)

The police personnel assigned to the division were given army ranks and uniforms (including the Reich cockade on the cap). While the ranks were retained,

Members of Headquarters, Police Division in 1939-40.

the division soon developed its own type of uniform: the police eagle reappeared on the caps, and the collar patches and belt buckle were also police issue; however, in keeping with Himmler's wishes, the national emblem was the model worn by the Waffen-SS and was worn on the upper right sleeve. Members of the SS also wore the embroidered version of the *Sig-Runen*[1] on the left breast pocket. Outwardly, about the only indications that the division was part of the army were its vehicle plates, which bore the WH code for *Wehrmacht-Heer* (Armed Forces-Army) instead of POL for police or SS for the Waffen-SS.

After three months of existence, in January 1940 the Police Division was moved from Wandern to the *Westwall* (Siegfried Line), where it relieved the 205th Infantry Division, part of XXV Army Corps, in the Kaiserstuhl area.

Shortly after the beginning of the Western Campaign the division was attached to the XVII Army Corps. As part of this corps, on 9 June 1940 the division began taking part in the campaign against France. Its strength at that time was:

Officers	NCOs	Enlisted Men	Total
400	1,999	31,162	33,561
1.2 %	6 %	92.8 %	100 %

This total obviously contained contingents for additional *Ordnungspolizei* battalions.[2]

The units first crossed the Ardennes Canal—which links the Meuse and the Aisne Rivers—southwest of Sedan and advanced toward Vouziers. On 14 June 1940 the troops reached Les Islettes (southwest of Verdun), having seen little of the enemy. East of Bar le Duc the unit crossed the Rhine-Marne Canal and pursued the retreating French through Neufchâteau and Besançon in the direction of the Swiss border. There, on 22 June 1940, word arrived of the ceasefire agreement. On 10 July 1940 the units were moved to the St. Dizier area as occupation troops, and on 2 August to the training camp at Suippes (50 km northwest of St. Dizier).

The SS Police Division
(*SS-Polizei-Division*)

At Suippes the division was reorganized and renamed the SS Police Division. The unit's strength was halved, mainly through the removal of men from older age classes. The army elements were relieved and replaced by newly-formed police units. In addition to the new Police Artillery Regiment and the Police Signals Battalion, a Police Reconnaissance Battalion was created from the former Police Bicycle Troop and a Police Flak Battalion established.

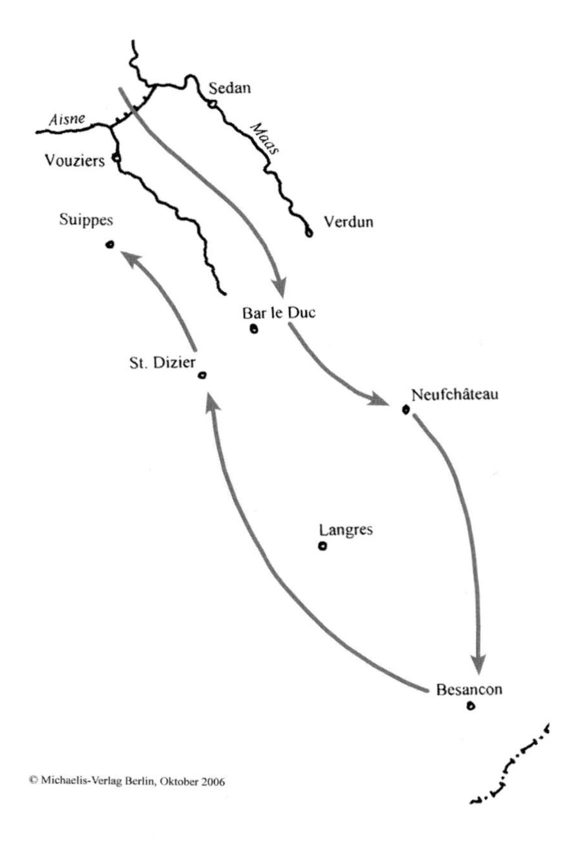

The Police Division in the French Campaign.

The commander of the Order Police Kurt Daluege during an inspection of the SS Police Division. Also present is division commander Mülverstedt.

The "SS" prefix was a sign of Himmler's desire, as *Reichsführer-SS* and Chief of the German Police, to demonstrate the close relationship that existed between the SS and the police.

On 12 November 1940 *Generalleutnant der Polizei* Mülverstedt assumed command of the SS Police Division. In mid-March 1941 *General* Halder advised that the unit would be taking part in "Operation Barbarossa." After discussions between *SS-Obergruppenführer* Daluege, *SS-Gruppenführer* Jüttner, and the *Reichsführer-SS*, on 17 April 1941 the SS Police Division was attached to the Waffen-SS with a strength of approximately 17,000 men. Two months later the order was received to move from France to L Army Corps in the east. The first troop trains arrived in East Prussia on 26 June 1941.

As had been the case during the campaign against France, when the Russian Campaign began the division was not deployed in the front line. It did not cross the Lithuanian border until 30 June 1941. Most of the roads were completely blocked, and so the march was a slow one in blazing summer heat. Within a few days the horse-drawn unit had lost about 600 of its 5,000 horses, and in Latvia it requisitioned horses or rented them along with their owners.

On 14 July 1941 the division crossed a pontoon bridge over the Dvinsk River near Dryssa. Two days later the artillery regiment became the first division unit to receive orders to go into action. On the night of 23-24 July the division, which still had not made contact with the enemy, marched through Ostrov. Beyond the town the troops entered an area of swamps and primeval forest. Attached to LVI Army Corps, on 1 August 1941 the division, under its new commander *SS-Brigadeführer und Generalmajor der Polizei* Walter Krüger, launched an attack to take the key position near Luga. The enemy had fortified the Luga position heavily, emplacing bunkers, mine fields, anti-tank barricades, and barbed wire in the complex terrain of forest, swamps, and lakes. It was also to the enemy's advantage that part of the area was a former army training camp. This extremely efficient defense inflicted heavy casualties on the SS Police Division. Attacks had to be called off and the troops withdrawn to their start lines. As Luga could not be taken through frontal attack, on 22 August 1941 the division moved southwest past Zopolye in order to attack from the flank. To avoid encirclement, on 1 September 1941 the Soviets evacuated the city. By then the SS Police Division had taken casualties of about 1,000 killed and more than 2,000 wounded. In those days the then *Gefreiter* Ochs wrote the following euphoric account:

"*Morning of 17/8/1941. Frozen stiff from the night cold, we lie in the foxholes hastily dug the previous evening. The morning fog slowly rises and from the east the light breaks through the dark of night. These are the best hours for us, for we love the light and feel much better and safer as soon as we can again see*

Summer 1941 – the advance on Leningrad. Below: abandoned Soviet 76-mm anti-aircraft guns, Model 1938.

the enemy, even if he is as strong as before. Word comes through from the left, `Unteroffizier Caspari to the company commander!' What's happening? He is my squad leader, therefore this interests me too. He has been my squad leader ever since I was transferred to the 11th Company. Throughout the wonderful time in Paris and the entire advance we have been together. But it has been combat that has welded us together. We know that he understands his business and we have unconditional faith in him.

Our fatigue has disappeared. We wait with interest for our squad leader to return. There he is. Tensely, we all look at him, but he is as cool as a cucumber. Acts as if it were nothing. But we know him, and we know very well that our squad has been given an assignment and that he is now quickly assessing the mission and making his decisions. He pulls out a map and studies it.

`Caspari Squad prepare for a reconnaissance in force mission.' That doesn't take us long, after all we're always ready for combat. We quickly grab extra ammunition, stick hand grenades into our boots, empty our pockets, and hand the letters, valuables, and documents they contain to the leader of the company headquarters squad. We make a quick but careful examination of our weapons and silently fall in behind our squad leader. We pass the foxholes of our comrades, their rifles lying within reach. There is no calling out to us, no, they greet and wish us success with their eyes. It is a brief brightening, but we know what it means.

Suddenly the familiar figure of Hauptmann Forstreuter, our company commander, appears from out of the mist. He is talking with two other officers. They are all studying maps they hold in their hands. Beyond our officers, about 15 meters away, is a group of pioneers: Feldwebel Rittweger and four men. They are all comrades we know from the previous fighting.

No sooner do we arrive when Leutnant Luger shouts, `Caspari squad and pioneers to me!' I have been ordered to lead a squad reinforced by pioneers to scout a favorable assembly area with good attack possibilities for a battalion attack on the barracks camp. You know how important the success of our mission is.'

He gives us a detailed briefing with groupings and distribution of forces. Then he says, 'The more silently and invisibly we can approach the enemy positions, the more we will be able to see and the greater will be our success.'

We move out in single file, with long intervals between men. The two men in the lead watch for mines. We swing far to the west, for yesterday's frontal attack against the Russian positions had to be called off because of the powerful defense system. We pass the combat outposts of the units deployed to our west, then come to a small stream. There we make a sharp change in direction and proceed straight north. Now we must be more cautious. Our movements become slower and more tentative. Artillery shells fall to our left and right, 100 to 200 meters away. We continue through a tall forest, whose undergrowth has been burnt away. Are we

Wounded are evacuated to the rear.

The division commander, Generalmajor der Polizei Mülverstedt.

walking into a trap? Why haven't we met any Russians? Or have we found an enemy weak spot on the first try? Suddenly we are standing in a clearing which stretches from west to east. The forest over there is burning! We have to get through. The light machine-gun goes into position and we quickly leave the forest. Once we reach the other side the machine-gunner automatically follows.

Half-burned bodies of fallen Russians lie in this forest. The Luftwaffe has paid a visit here. Huge bomb craters, beautifully laid out, accompany us. We must be getting close to the barracks camp, or at least to the sand pit to its southwest. Then a signal from the front: 'Get down!' We look tensely to see if we can determine the reason for the halt. To the left-front something white shimmers through the tree trunks—it's the sand pit. Slowly we work our way towards it, keeping a careful watch to all sides. We immediately see that we should have a good all-round view from a place where the sand pit rises and there are no trees.

We reach the edge and before us see a large sandy area. In the middle is a wooden hut, which is burning. Someone must have been there recently. In the midst of these contemplations Leutnant Luger waves. By now he has reached an elevated position. Down below, as shown on the map, there is a brook and the ground is swampy. Then the terrain rises sharply and opens onto a plateau. On it is tomorrow's objective, the barracks camp. Suddenly a machine-gun opens up to our right. The battle noises grow increasingly louder and more hectic. Then Unteroffizier Caspari says to the Leutnant, 'Herr Leutnant, we have completed our mission, it would be best to head back now.' 'Not quite Unteroffizier Caspari. I have the assembly area but not the route for the subsequent attack. Feldwebel Rittweger, will you come with me?' 'Of course, Herr Leutnant.' 'Then I still need volunteers, preferably unmarried riflemen.'

Everyone volunteered. Leutnant Luger took Obergefreiter Lisson and Gefreiter Ochs. 'Uffz. Caspari, you stay here with the rest of the men and cover us in case we're spotted and have to retreat. Don't leave this spot until we return.'

Carefully we climb into the stream bed. Through the light undergrowth we can easily see the Russians standing beside their positions or disappearing into them. Perhaps they think we're still south of their positions and consider an attack from this side impossible because of the swampy terrain.

Our movements become ever more cautious. We check the ground carefully before putting a foot down. In front of us the Russian sentries are close enough to touch.

Then Leutnant Luger gives us hand signals indicating that he wants to go on alone with Feldwebel Rittweger. We are to guard against surprises as they press ahead. Tensely we watch our two leaders. Will it succeed? They have been gone for some time. We become restless. Obergefreiter Lisson wants to leave the position to see if he can catch sight of them. No, there's no point, for if the two are spotted and have to fall back there'll be no covering fire. Finally the pair returns. They're sweaty and their faces are red, but their eyes are shining. The mission has been accomplished. The battalion will attack and take the barracks camp as it states in the regiment's order."

The division was again sent to L Army Corps, and on 9 September 1941 the battle began for Krasnogvardeysk, one of the main blocking positions in front of Leningrad. Five days later the SS Police Division took the city. Once referred to as "The Guard Division of Paris," the unit displayed a good fighting strength. Hitler halted the attack on Leningrad, however, so that he would not have to feed the civilian population during the winter.

On 18 October 1941 the division received 3,000 replacements to make good its losses. In the weeks that followed, however, there were ever more serious shortages. Not only did the artillery remain without shells, but the soldiers also lacked winter clothing and food. The division remained in the siege ring around Leningrad, in the Pushkin – Pulkovo area, until February 1942, now under the command of *SS-Brigadeführer und Generalmajor der Polizei* Wünnenberg.

On 13 January 1942 the Red Army broke through at the Volkhov River about 40 km north of Novgorod, at the junction between the 16th and 18th Armies. Hitler ordered the SS Police Division committed at the Volkhov. On 19 February 1942 the division handed its positions in the Leningrad siege line over to the 121st Infantry Division, leaving the artillery regiment behind, and marched into the Chudovo area. Attached to I Army Corps, the unit—which had been absorbed into the SS on 10 February 1942[3]—was supposed to restore contact with XXXVIII Army Corps. The SS Police Division attacked on 15 March 1942 and four days later closed the so-called Volkhov Pocket.

The Red Army made repeated determined attempts to break open the pocket from within and without, resulting in fierce fighting. This, combined with the terrible winter conditions with temperatures as low as -50°C, caused the combined strength of the three rifle regiments to fall to about 500 men by the end of April 1942. The *Wehrmacht* communiqué reported:

"*In the northern sector, in several days of offensive operations in extremely adverse terrain and weather conditions, units of the army and the Waffen-SS counterattacked powerful enemy forces that had broken into their positions and drove them back. The SS Police Division, in particular, distinguished itself in the fighting.*"

Outside Leningrad…

The pocket in this region, with its primeval-like forests, was finally eliminated at the end of June 1942, and the division was again attached to the L Army Corps and moved north into the Sablino area.

From there the unit again moved to the Leningrad front and occupied positions near Krasny Bor. The Red Army launched a counterattack on 23 July 1942. After the costly defensive fighting died down, on 8 August one battalion from each regiment was ordered to SS Training Camp "Heidelager." For the troops left in the main line of resistance, one week later there was another major Soviet attack in the Kolpino area. On 20 August 1942 the division received 2,000 replacements to make good its losses.

When Headquarters, 11th Army was moved from the Ukraine to Leningrad to take the Soviet metropolis in September, the Red Army renewed its offensive. There was extremely heavy fighting until the beginning of October 1942, and the German troops again succeeded in destroying powerful enemy forces.

In mid-October 1942 the SS Police Rifle Regiments were renamed SS Police Infantry Regiments and the SS Police Reconnaissance Battalion became the SS Police Bicycle Battalion.

In November 1942, with three infantry regiments of two battalions each, the unit reported its strength as follows:

Officers	Other Ranks	Total
322	10,157	10,479
3.1%	96.9%	100%

In addition, the three new infantry battalions being formed at SS Training Camp Heidelager reported:

Officers	Other Ranks	Total
48	2,872	2,920
1.6%	98.4%	100%

With about 13,400 men, the SS Police Division thus had approximately 80% of its authorized strength of about 16,500 men.

From left to right: SS-Hauptsturmführer Traupe, SS-Sturmbannführer Schüpers, SS-Untersturmführer.

Division commander SS-Brigadeführer Wünnenberg awards the German Cross in Gold to SS-Untersturmführer Butte on 18/9/42.

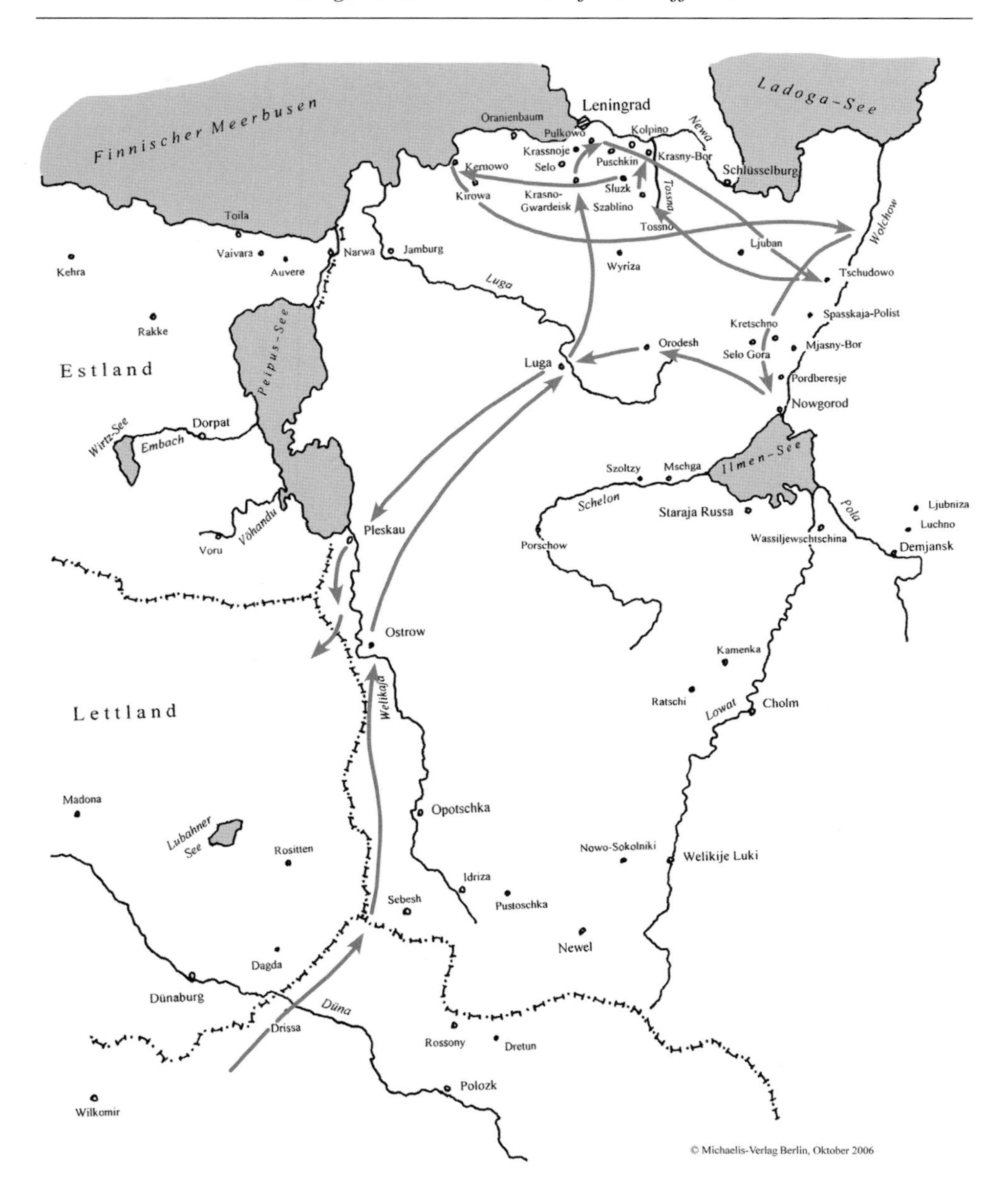
Finnischer Meerbusen
Ladoga-See
Leningrad
Oranienbaum
Pulkowo
Kolpino
Newa
Krassnoje
Selo
Puschkin
Krasny-Bor
Schlüsselburg
Kemowo
Kirowa
Krasno-
Gwardeisk
Sluzk
Szablino
Tossna
Tossno
Wolchow
Toila
Vaivara
Auvere
Kehra
Narwa
Jamburg
Luga
Wyriza
Ljuban
Tschudowo
Spasskaja-Polist
Kretschno
Selo Gora
Mjasny-Bor
Orodesh
Rakke
Peipus-See
Estland
Pordberesje
Nowgorod
Wirtz-See
Embach
Dorpat
Szoltzy
Mschga
Ilmen-See
Schelon
Staraja Russa
Pola
Ljubniza
Luchno
Demjansk
Wassiljewschtschina
Porschow
Vöhandu
Voru
Pleskau
Ostrow
Kamenka
Ratschi
Lowat
Cholm
Lettland
Welikaja
Madona
Opotschka
Lubahner See
Rositten
Nowo-Sokolniki
Welikije Luki
Idriza
Pustoschka
Sebesh
Newel
Dagda
Dünaburg
Düna
Drissa
Rossony
Dretun
Polozk
Wilkomir

On 12 January 1943 the Soviets launched their winter offensive on the northern front. Four days after enemy troops succeeded in penetrating the front of the XXVI Army Corps to the south, the SS Police Division received orders to reestablish contact with Schlisselburg, free the troops encircled there, and simultaneously restore the ring around Leningrad. The German troops had no answer to the enemy's tremendous numbers in tanks and artillery. Schlisselburg had to be abandoned. The SS Police Division was almost wiped out in the fierce fighting in the Sinyavino area. Battalion combat strengths fell to just 30 to 50 men. When the fighting rose to an inferno pitch on 28 January 1943 some troops fled before the enemy in panic. The next day what was left of the division was pulled out of the line and moved back to LIV Army Corps. In Sablino three weak battalions were created from the three infantry regiments—designated SS police panzer-grenadier regiments since 1 February 1943. Every cook, laborer, horse tender, and clerk was forced to join the fighting units. In the recent fighting the SS Police Division had taken approximately 3,000 casualties.

On 9 February 1943 a newly-formed battalion (1,200 men) arrived from SS Training Camp "Heidelager." The other two battalions had already been sent to the southern front at the end of 1942.[4]

On 10 February 1943 the Red Army resumed its offensive, but by scraping together all available forces, the German troops succeeded for the most part in holding the areas. The fighting died down at the end of February 1943 and remained so along the Tosna River until mid-March.

South of Kolpino there was no peace for the grenadiers, already hard pressed by the terrible weather conditions. On 18 March 1943, after massing tremendous numbers of men and guns, the enemy attacked again. The front held, and after 14 days the Red Army halted its attacks. For about ½ of the SS Police Division this meant finally being taken out of the front during April. Some elements, especially the cadres for the reorganization of the unit as a panzer grenadier division, were sent to SS Training Camp "Heidelager." Left at the front were:

SS Police Artillery Regiment
SS Police Flak Battalion
III Battalion/SS Police Grenadier Regiment 1
I Battalion/SS Police Grenadier Regiment 2
I Battalion/SS Police Grenadier Regiment 3

Bolstered with additions from other units, under the command of *SS-Standartenführer* Bock these elements formed Battle Group/SS Police Division in

37-mm Pak 35/36 in the area north of Novgorod.

the siege ring around Leningrad. Initially there was no further fighting there. On 21 October 1943 the three battalions together were designated SS Police Panzer Grenadier Regiment 3.

In November 1943 elements of the 24th and 225th Infantry Divisions took over the positions of Battle Group/SS Police Division, which moved to the western front of the Oranienbaum pocket between Kernovo and Gorbovitsy—attached to L Army Corps. When the III SS Panzer Corps arrived from Croatia at the end of 1943 and took over the area, the SS battle group came under this new command with a strength of:

Officers	NCOs	Enlisted Men	Total
111	903	4,053	5,067
2.2%	17.8%	80%	100%

When the main body of the corps arrived at the Oranienbaum pocket, on 10 January 1944 the Battle Group/SS Police Division was pulled out of the front and moved almost 200 km to the east to the Volkhov. Attached to XXVIII Army Corps, it relieved elements of the 96th Infantry Division and occupied positions at the mouth of the Tigoda River.

On 16 January 1944, after the Soviets launched their new winter offensive, heavy fighting broke out there. The focal point of the attack was farther south near Novgorod, however, and just five days later the unit, now designated SS Battle Group Bock, was pulled out of the main line of resistance and ordered to the XXXVIII Army Corps. In the fighting that followed the battle group was almost completely destroyed by the Red Army. The survivors fell back through Sapalye to Oredesh, losing most of their heavy weapons. Together with Latvian SS volunteers, they defended the town until they were encircled. A breakout was made at the cost of heavy casualties and the survivors made their way via Luga—which had fallen to the SS Police Division on 24 August 1941—to Turkoviche, where they reached the German main line of resistance.

With no hope of offering organized resistance the troops streamed back in the direction of Pskov, and on 2 March 1944 occupied provisional positions in the Panther Position. Two days later the Red Army attacked the desolate units along the Velikaya. With no supporting heavy weapons, the final action by SS Battle Group Bock was defending against Soviet assaults between Pskov and Ostrov on Lake Pskov. Finally acknowledged as no longer capable of operations, at the end of March the remnants of the unit were sent to SS Training Camp "Kurmark" by way of Opochka.

Besitzzeugnis

Dem SS-Oberscharführer
(Dienstgrad)

Siegfried Höcke
(Vor- und Zuname)

I./SS-Panzer-Grenadier-Regiment 7
(Truppenteil)

verleihe ich das

Infanterie-Sturmabzeichen

— Silber —

Rgt.St.Qu., den 30. 1. 1944
(Ort und Datum)

Nr. 2956/44

(Unterschrift)

SS-Obersturmbannführer
und Regiments-Kommandeur
(Dienstgrad und Dienststellung)

SS-Oberscharführer Höcke was awarded the Infantry Assault Badge in Silver after the fighting near Novgorod.a

Grenadiers of the SS Police Division at the Volkhov.

The 4th SS Police Panzer Grenadier Division
(*4. SS-Polizei-Panzergrenadier-Division*)

The reorganization of the infantry division, still reliant on horses for transport, into a panzer grenadier division began in April 1943. Because of the situation at the front, however, only about 2,500 men were transferred to SS Training Camp "Heidelager" or to schools for specialized training (pioneers and signals personnel, for example). The division was to be brought up to authorized strength (see below) mainly through the addition of ethnic Germans from Rumania.

Officers	NCOs	Enlisted Men	Total
520	3,355	13,130	17,005
3.1%	19.7%	77.2%	100%

Formation of the frameworks of the new units began while SS Battle Group Bock was still fighting with Army Group North. The new SS police panzer grenadier division was to be organized as follows:

Division Headquarters
SS Police Panzer Grenadier Regiment 1 (I – III Battalions)
SS Police Panzer Grenadier Regiment 2 (I – III Battalions)
SS Police Panzer Grenadier Regiment 3 (I – III Battalions)
SS Police Artillery Regiment
SS Police Flak Battalion
SS Police Armored Battalion
SS Police Signals Battalion
SS Police Armored Reconnaissance Battalion[5]
SS Police Anti-Tank Battalion
SS Police Pioneer Battalion
SS Police Economic Battalion

SS Police Panzer Grenadier Regiment 3 was not a new unit created at the training camp; instead, the infantry units of SS Battle Group Bock formed the third grenadier regiment.

In April 1943 the first units were sent from SS Training Camp "Heidelager" to the *Generalgouvernement* for action against Polish partisans. Then in July 1943 Mussolini was removed from power. In response to the anticipated changing of sides by the Italians German forces had to take over large areas of the Balkans, and the OKW ordered elements of the SS Police Panzer Grenadier Division to the 2nd Army.

At the end of July 1943, SS Police Panzer Grenadier Regiment 1 moved into quarters in the Valjevo area southwest of Belgrade and SS Police Panzer Grenadier Regiment 2 in the Misar/Sabac area (70 km west of Belgrade). When, in the course of the new political situation, the Italian III Army Corps was ordered home from eastern Greece, the two panzer grenadier regiments were ordered to move to the new area of operations. Attached to XXII Mountain Corps along with the 1st Mountain and 104th Light Infantry Divisions, the units were supposed to continue training while taking on security duties against the Greek ELAS and EDES partisans.

A large-scale, active anti-partisan campaign was hardly possible. In fact, in the Larissa area (bordered by the Vistritsa Mountains in the north, the Pindus Mountains in the west, the Gulf of Salonika in the east, and the Gulf of Volos in the south) only road and rail communications were guarded by strongpoints (earth bunkers with barbed wire and mine fields). The German occupation troops were frequently attacked by Greek partisans and, after the arrival of the SS Police Armored Battalion in the spring of 1944, assault guns were deployed at the smaller strongpoints. A former member recalled:

"*We detrained in Larissa, then spent the first night at the small station. The next day we were issued 40 mules. Senior NCO Schade had a lot of trouble finding suitable people for them. Under the command of SS-Unterscharführer Gonschauer, the mules were then brought to Ellasong in two days. The town lay at the foot of Mount Olympus. We were supposed to guard the Winkler Pass there. One platoon was to patrol the mountains daily on mules, another was to patrol the pass in vehicles, and the last platoon was to complete its training in the town. The duties were exchanged daily. The 1st Company had its first tough day in Greece on 10 October 1944. It was patrolling a side road up in the high mountains. Everyone was in good spirits, but suddenly there was a tremendous explosion in front of us. They had blown up the roadway right under our noses. With picks and shovels*

SS-Brigadeführer Fritz Schmedes commanded the 4th SS Police Panzer Grenadier Division, with interruptions, from June 1942 to November 1944.

Walter Harzer led the division until the end of the war with the rank of SS-Standartenführer.

we soon made the road passable again. This was repeated a few times. Then SS-Untersturmführer Dirol made a serious mistake. He had us drive on without a screening force. We were in a basin when the firing started. Our vehicles were riddled by small arms fire in no time and rendered unusable. They were all write-offs. The heavy firing went on without letup from half-past ten in the morning until dark. We had 11 killed and 23 wounded. Things quieted down when darkness fell. The men were organized into three march groups, taking the wounded with them. The next morning we had to fight our way out. The fallen were recovered on 12 October 1944. They were laid out in Ellasong and buried with military honors. Later there was a sequel to this incident. SS-Untersturmführer Dirol had falsified the report on the Ellasong battle, resulting in an SS-Unterscharführer being sentenced to death—he was later cleared. Dirol was brought before a court martial in Larissa and was sent to the SS and Police Prison Camp at Danzig-Matzkau."

Hygienic conditions were a complicating factor throughout the entire period of occupation duty. Approximately 80% of the soldiers fell sick with malaria or developed dysentery—mainly because of unclean water. Another limiting factor was the incomplete training of the young recruits, which limited their operational capabilities.

As part of the renumbering of Waffen-SS units, effective 22 October 1943 the division—which was attached to Senior Administrative Headquarters 395 (commander Salonika – Aegean)—was given the designation 4th SS Police Panzer Grenadier Division.

While the division units were also identified by the number 4, the panzer grenadier regiments were given the numbers 7 and 8.[6] The word "Police" was dropped from all unit titles except for that of the division. On 31 December 1943 the division reported the following strength (not including the battle group at the front):

Officers	NCOs	Enlisted Men	Total
216	1,399	9,094	10,709
2%	13%	85%	100%

Summing up less than a year after formation of the motorized SS division began, its new commander *SS-Brigadeführer und Generalmajor der Waffen-SS und Polizei* Schmedes declared that unit training could not be considered complete because of missing weapons and equipment and lack of fuel. As well, occupation duties made the tactical employment of the entire division impossible. The units were almost all deployed separately in the area.

On 23 May 1944 Army Group E ordered elements of the 4th SS Police Panzer Grenadier Division to relieve SS Police Mountain Regiment 18 on the Gulf of Corinth. Together with I Battalion, SS Artillery Regiment 4, SS Panzer Grenadier Regiment 7 subsequently moved the approximately 120 kilometers to the south. The security and pacification duties there resulted in a considerable upsurge in violence. In the village of Distomon—a few kilometers from the coast—a number

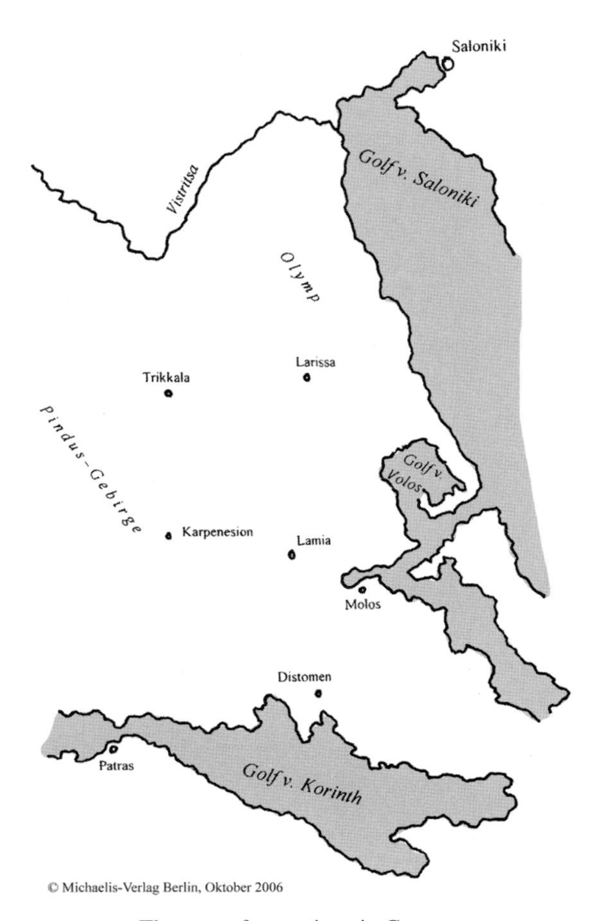

The area of operations in Greece.

Franz Poroda (age class 1909) was killed near Klein-Betschkerek (Temeschburg District) on 23/9/44. While on leave in 1943 to help with the sugar beet harvest, he described the reprisals in Greece, which had affected him deeply.

of Greek civilians were executed as a reprisal, resulting in a court martial inquiry against the responsible company and battalion commanders.

The bulk of the former SS Battle Group Bock was sent from SS Training Camp "Kurmark" to Greece, joining the other units already there. On 30 June 1944 the division submitted the following strength report:

Officers	NCOs	Enlisted Men	Total
385	2,491	13,015	15,891
2.4%	15.7%	81.9%	100%

Authorized strength at that time was:

Officers	NCOs	Enlisted Men	Total
557	3,388	12,595	16,538
3.4%	20.5%	76.1%	100%

With proper training the unit had the potential to become a powerful unit. Of the personnel in the units, about 50% had experience in Russia, while the rest had a year's experience as occupation troops.

In July 1944 elements of the two SS panzer grenadier regiments were ordered to clear an avenue of retreat for the LXVIII Army Corps to the south in the southern Pindus Mountains. At the end of August 1944 the situation took a dramatic turn for the worse when Rumania changed sides. To shore up the endangered front and prevent the German forces in the south from being cut off, the 4th SS Police Panzer Grenadier Division was ordered to move to Werschetz, near Belgrade, about 600 kilometers to the north in Army Group South Ukraine's area.

While the division was being transported from Larissa via Salonika to Serbia, Bulgaria also declared war on Germany. Elements of the division were subsequently employed in the Skoplĕ (Kosovo) area to disarm Bulgarian troops, previously occupation forces. While one battle group saw action with the 21st SS Waffen Grenadier Division "Skanderberg" (Albanian No. 1), the bulk of the 4th SS Police Panzer Grenadier Division guarded important lines of communication from Serbia to Kosovo.

Divided into battle groups, the division—moving mainly by night due to the threat of air attack—began arriving in the Belgrade area on 11 September 1944. The retreat road was littered with burnt out vehicles and wrecked wagons with dead horses.

On arriving in Belgrade, the unit's vehicles had to be overhauled after their strenuous journey. The tires, in particular, were badly worn and had to be replaced. The assault guns had been left behind in Salonika and did not join the units until later. These left Belgrade in pouring rain, crossed the Danube by ferry near Pančevo, and from there were transported by rail to the Temesvar area in the Rumanian Banat.

Attached to LVII Panzer Corps, the 4th SS Police Panzer Grenadier Division was supposed to retake Temesvar (German Temeschburg), which had fallen to the Red Army. Because of the situation *SS-Brigadeführer* Schmedes refused to carry out the attack and instead had his unit, which was under constant enemy pressure, occupy a bridgehead across the Theiss near Szeged. The Chief of the Army General Staff subsequently sent a telex to Army Group South:

"*The 4th SS Police Panzer Grenadier Division's anticipated behavior is not in keeping with the Führer's orders. I hereby order you to instruct the division to bypass the city of Temeschburg to the south and east and take possession of and barricade the road from the Iron Gate through Caransebes to Temeschburg.*"

As *SS-Brigadeführer* Schmedes, whom Himmler described as a "weak character," refused to carry out the order, soon afterwards he was relieved by *SS-Standartenführer* Harzer. Under continued heavy pressure from the Red Army, the division moved approximately 100 kilometers into the Szolnok area, evacuating ethnic Germans as it went. A former member of the division recalled:

"*In the afternoon we drove on in the direction of Perjamosch, which was already occupied by Russian troops. Coming from the south, the Pesak Road opened into the southern edge of the town west of the station. At the head of our column were three of the regiment's assault guns. Behind them were the trucks with the panzer grenadiers. The column halted about 2 kilometers from Perjamosch. We got down from the trucks and formed up for an attack. One of the assault guns stayed on the road, while the other two deployed 50 meters to the left and right in preparation for an advance on Perjamosch parallel to the road. The panzer grenadiers, at least 400 men of I Battalion, 8th Regiment, spread themselves out in platoons and squads around the assault guns. Several 75-mm anti-tank guns, 20-mm anti-aircraft guns, and infantry guns took up position in front of Perjamosch. Also there was one of the company's mortar platoons. When the attack began, the middle assault gun was engaged by a Soviet anti-tank gun, and the attack on the road was somewhat held up. As a result the assault gun on the right, behind which I was running, got too far forward. The Soviet anti-tank gun was thus able to hit our assault gun with two shells and set it on fire. The assault gun crew abandoned their vehicle. Miraculously I was not hurt, even though red-hot fragments were flying all around me. At that moment one or two enemy machine-guns opened fire. At that time I was about 15 meters behind the burning assault gun. Lying on my belly, I crawled slowly to the rear, as I knew that the burning assault gun would soon blow up.*

Meanwhile, the other two assault guns were able to move up. From the Perjamosch station we then charged in the direction of the Rumanian church, where we suspected the forward observer was. On the right, my heavy machine-gun squad stormed toward the cemetery. There we were met by heavy Soviet fire. We were able to take out a heavy machine-gun position with hand grenades. The Russians fled. In the evening I searched several houses near the power plant for cellar entrances, as I suspected there were civilians there. In one house I discovered a hidden trap door leading from the dining room to the cellar. About 50 Banat ethnic Germans were hiding from the Soviets there, and they were surprised to discover we were Germans. As we were all hungry, we were given food and drink, in keeping with the hospitality shown everywhere in the Banat.

We were able to carry out an orderly evacuation of Perjamosch. The civilians headed west on tractors and their trailers, horse-drawn vehicles, etc. For the next few days the Soviets made repeated attempts to reenter Perjamosch but were repulsed each time.

It was a beautiful Sunday and everything was quiet. My comrades and I had time to wash and search for the lice we had brought with us from Greece. We had no clean clothes and were still wearing the light denim camouflage uniforms we had worn in the heat in Greece. In the meantime the mess tins had arrived filled with food, and each man was peacefully engaged in eating his meal. Suddenly we heard the Russian 'Urray!' battle cry and saw about 100 Russians charging towards us. The 1st and 2nd gunners leapt to the machine-gun, however, it refused to give continuous fire. In seconds the Russians had almost crossed the meadow. In this situation the leader of the heavy machine-gun squad and I each grabbed a 98k carbine and from the barn window began shooting the closest Russians, one after another. When the following Russians realized that the men closest to the road were being knocked down they began lying down in the meadow beyond the road. Two carbines and a crippled heavy machine-gun had prevented about 100 Russians from crossing the road and overrunning us. A few minutes later, after the breech and barrel had been replaced, the heavy machine-gun began firing. By then a squad of reinforcements had also reached the company command post. Half an hour later there were only dead Russians in the meadow before us. For me and my comrades from the Banat and Transylvania, at the time it had simply been a matter of doing our duty, defending our Banat homeland! Today I feel bitter that I, just a 19-year-old boy, had been misused as a tool of the ideological struggle.

On 21 October 1944 I was again a forward lookout in front of the main line of resistance. I and my friend Matz from Friedburg, near Temeschburg, were each in a foxhole watching for the Soviets. As there was no movement to be seen, in beautiful sunshine I walked over to an isolated Hungarian farmhouse that lay in no-man's-land about 500 meters to our left front. On the way I picked up four hens that were running loose around our foxholes. As I speak Hungarian, I asked the Hungarian farm family to make us a proper Hungarian paprika chicken. They told me that there was a dead German soldier 500 meters farther on, off to the right. He had been lying beside a country road for tens days. It was then that I first became aware that this ground had changed hands between the Germans and Russians several times in recent days. Everywhere we found completed foxholes, the excrement of their former owners, dead animals—horses, for example—and dead people (civilians, Red Army, German). The bodies were all decomposing and stank terribly. As the area was quiet, I took two of the family's sons with me with spades. When we came to the dead man I could recognize his German uniform and saw that he had blonde hair. Maggots covered his face and body and the smell was terrible. The two boys immediately began digging a hole beside him, and I tried with two small sticks to pluck the German identity disc out of the maggots and rotten flesh on his chest. I grasped the disc in a piece of cloth, broke it in two, and kept one half. Using a pencil, I copied the characters that were on the identity disc onto a piece of wood. I then placed the piece of wood atop the burial mound where it could be seen. By then it was late afternoon. I sent the boys home and began walking in the direction of my forward outpost. No sooner had I started walking when an Unterscharführer came up to me. He told me I should hurry, because our artillery was about to begin ranging in on this area. I telephoned the company command post and reported the burial beyond the main line of resistance. Later a motorcycle-sidecar came and collected me, and I had to mark on a general staff map exactly where I had buried the man. A report was written, and the half of the recognition disc I had retrieved was attached to it. The next day I was wounded by shrapnel from a mortar round, and finally, on 9 December 1944, my leg had to be amputated above the knee! Afterwards I weighed less than 37 kg!"

The Soviet advance on Budapest was supposed to be halted beyond the Theiss (Tisza); however, on 24 October 1944 the Red Army succeeded in crossing the river. After fierce fighting, 4th SS Police Panzer Grenadier Division was forced to

Im Namen des Führers
und Obersten Befehlshabers
der Wehrmacht

verleihe ich

dem

Oberleutnant in der SS-Pol.-Nachrichten-Abt.

Werner Schmoll

das

Eiserne Kreuz 2. Klasse

Div.Gef.St., den 30. Okt. 1941

Nr. 1578/41

SS-Brigadeführer
Generalmajor der Waffen-SS u.
Divisions-Kommandeur

(Dienstgrad und Dienststellung)

On 30/10/41 SS-Brigadeführer Krüger awarded the Iron Cross, Second Class to Oberleutnant Schmoll in the Pushkin area outside Leningrad.

IM NAMEN DES FÜHRERS
UND
OBERSTEN BEFEHLSHABERS
DER WEHRMACHT
IST DEM

SS-Hstuf. Werner S c h m o l l
2./SS-Nachr.Abt.4

AM 1. 8. 1942

DIE MEDAILLE
WINTERSCHLACHT IM OSTEN
1941/42
(OSTMEDAILLE)
VERLIEHEN WORDEN.

FÜR DIE RICHTIGKEIT:

SS-Sturmbannführer u. Kdr.

The so-called "Frozen Flesh Medal" awarded to Schmoll for actions outside Leningrad and at the Volkhov River.

IM NAMEN DES FÜHRERS

VERLEIHE ICH

DEM

SS-Hauptsturmführer

Werner S c h m o l l

2./SS-Nachrichtenabt.4

DAS

KRIEGSVERDIENSTKREUZ

2. KLASSE

MIT SCHWERTERN

Div.St.Qu. , DEN 20.4.1944

Nr. 587/44

SS-Brigadeführer

und Generalmajor der Waffen-SS

(DIENSTGRAD UND DIENSTSTELLUNG)

Div.-Kommandeur

SS-Brigadeführer Schmedes awarded the War Merit Cross 2nd Class with Swords to SS-Hauptsturmführer Schmoll in Greece on 20/4/44.

Im Namen des Führers und Obersten Befehlshabers der Wehrmacht

verleihe ich

dem

ᛋᛋ-Hauptsturmführer

Werner Schmoll,

ᛋᛋ-Nachr.Abt. 4

das

Eiserne Kreuz 1. Klasse.

..Div.Gef.St......., den ..9..Dezember..19..44
Nr.1084/44

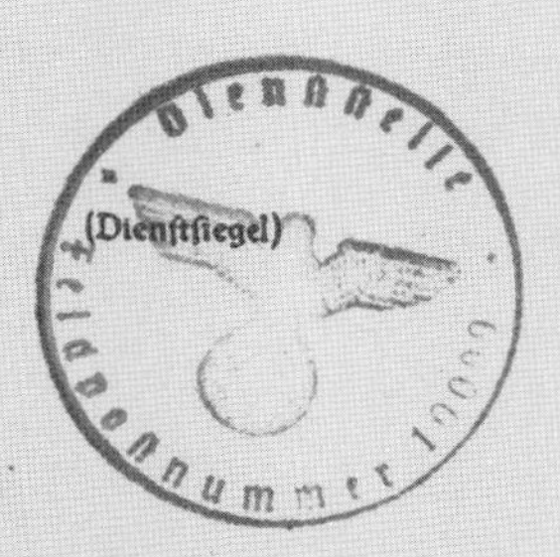

(Dienstsiegel)

ᛋᛋ-Brigadeführer
und Generalmajor der Waffen-ᛋᛋ
Div.-Kommandeur
(Dienstgrad und Dienststellung)

SS-Hauptsturmführer Schmoll was awarded the Iron Cross, First Class for a special act of bravery during the defense of Budapest.

withdrawn into the Jász-Ladany area. There the 18th SS Volunteer Panzer Grenadier Division "Horst Wessel" was attached to the unit. The division proved a complete failure in the fighting on 11 November 1944. Of this, Army Group South wrote:

"In the sector of the 4th SS Police Panzer Grenadier Division attached to LVII Panzer Corps, elements of the 18th SS Division were a complete failure, allowing themselves to be overrun and surrendering."

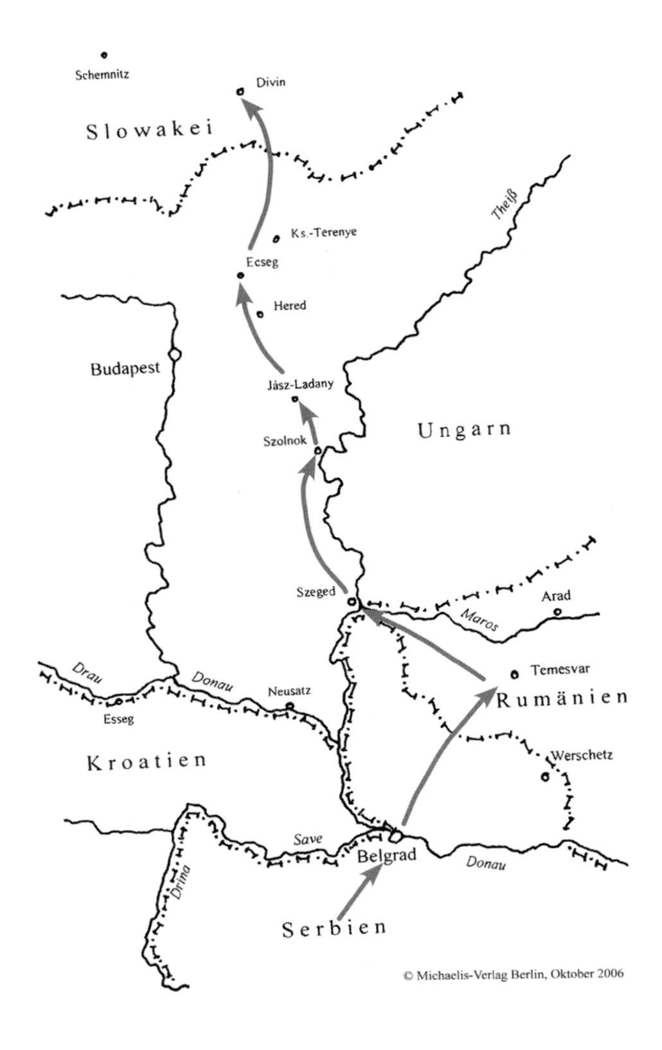

The retreat from Serbia into Slovakia.

The units withdrew via Heréd toward Ecséd, which was reached on 25 November. There the division was at least able to address some of its losses in personnel, receiving replacements from SS Field Replacement Battalion 4.

For a long time the German troops had only been capable of reacting to Soviet moves, and in the subsequent fighting the division, under the command of *SS-Standartenführer* Harzer, was almost encircled south of Kis-Terenye. Attached to the Panzer Corps "*Feldherrnhalle*," the division moved into the Slovakian-Hungarian border region, and in mid-January 1945 it was ordered to seal off a Soviet breakthrough in the Divin area (about 20 kilometers from the border).

In the days that followed, the elements of the 18th SS Volunteer panzer grenadier division attached to the unit were released. Having lost much of its heavy equipment during the retreat from Greece, the 4th SS Police Panzer Grenadier Division took over the equipment of the attached units. As well I Battalion, SS Volunteer Panzer Grenadier Regiment 40 was incorporated into the division.

In the Schemnitz area at the end of January 1945 the division received orders to move to Pomerania, where Soviet forces were already advancing toward Berlin along the Netze and Warthe Rivers. The first element of the division, SS Panzer Grenadier Regiment 8, arrived in Stargard on 1 February 1945 and was immediately committed southeast of Greifenhagen. The entire division assembled there—between the Oder and Lake Madü—during the first weeks of February.

In preparation for the German offensive in Pomerania, which was supposed to take the Russians in the flank from the north, the 4th SS Police Panzer Grenadier Division moved into the area south of Stargard, where it came under the command of XXXIX Panzer Corps. Fighting began on 16 February 1945, and with SS Panzer Battalion 4, which was again battle-ready, the unit was able to advance toward Dölitz. In the face of superior Soviet forces the offensive was halted on 19 February 1945. The next day the 10th SS Panzer Division took over the 4th SS Police Panzer Grenadier Division's positions. The latter was ordered to "*take over the defense of Danzig*" because "*it is still very well-equipped.*"

The division entrained on 25 February 1945 and was transported to VII Panzer Corps in the Dirschau area. The 4th SS Police Panzer Grenadier Division was unloaded in the Rummelsburg area short of its destination. There the VII Panzer Corps was supposed to reestablish contact with the 3rd Panzer Army. The 4th SS Police Panzer Grenadier Division launched a counterattack on 28 February and was at least able to provide some relief. Soviet pressure forced the German units to retreat steadily, however.

To prevent the German main line of resistance from being rolled up along the coast to Danzig, the 4th SS Police Panzer Grenadier Division moved into the Stolp area and withdrew through Lauenburg to Neustadt. The roads were completely clogged with refugees and Wehrmacht vehicles. Soviet aircraft and tanks shot up many such columns.

Sturmgeschütz IV (Sd.Kfz. 163) of SS Panzer Battalion 4.

Spring 1945.

On 9 May 1945 the Commander-in-Chief of the 2nd Army assessed the units under his command. He characterized the 4th SS Police Panzer Grenadier Division as follows:

"*Battle-tested division, with new replacements lacking combat experience... Heavy casualties, having been in the thick of the defensive fighting. Rations strength: 4,767, fighting strength: 2,744.*"

At that time the division had left:

1 Panzer IV
8 Sturmgeschütze III
6 Sturmgeschütze IV
11 75-mm anti-tank guns (towed)

With the combined strength of the division's units reduced to that of a strong regiment, they were subsequently referred to as Battle Group/4th SS Police Panzer Grenadier Division. Consisting of 5 battalions, 1 pioneer battalion, 1 field replacement battalion, and three attached companies of French SS members,[7] the unit took over the area from the Putziger Wiek (Puck Bay) to Rahmel.

After sometimes fierce fighting, on 12 March 1945 the Soviets took Neustadt, unleashing a desperate flight by German soldiers and civilians in the direction of Oxhöfter Kämpe and Gotenhafen. VII Panzer Corps was able to temporarily stabilize the front there. The bulk of the rear-echelon units, civilians, and vehicles were massed on the Kämpe in such a way that the enemy's massed artillery fire caused extremely heavy casualties.

On 20 March 1945 the Red Army split the Danzig – Gotenhafen defense zone. Four days later Hitler ordered that Hela, Gotenhafen, Danzig, and Königsberg, all of which had been declared "fortresses," were to be held at all costs. The order was soon overtaken by events, however, for on 28 March 1945 the Red Army took Gotenhafen. The situation north of the city on the Oxhöfter Kämpe deteriorated steadily. To save the approximately 20,000 men of VII Panzer Corps and elements of the civilian population from destruction or captivity, on 4 April 1945 *General der Panzertruppen* Kessel, the commanding general, ordered a withdrawal to Hela. During the evacuation across the Putziger Wiek all of the remaining vehicles were destroyed at the Oxhöfter Kämpe airfield. In a masterful feat, by the next day the

Kriegsmarine had evacuated all of the German soldiers and many civilians to the peninsula. A former member of the division recalled:

"During the fighting in Danzig a Feldwebel from the radio company came and told us that he had become separated from his unit along with his equipment. Signals equipment, especially complete with a vehicle, was something we could use. After questioning him, we learned that he and the radio set belonged to a propaganda unit and that the company had been completely scattered in the fighting. We were a little uncertain about using the equipment for normal radio communications, however, for our most powerful transmitter was a 100-Watt set and this one was 200-Watts. It could be heard from a long distance, and even at the end of the war we tried to follow the manuals as best we could.

It wasn't long before the Danzig pocket was closed. An entire corps was encircled. Using the 200-Watt transmitter, we finally established contact with Führer Headquarters. The entire corps now depended on this link, and not long afterwards we had to send a message to Berlin describing the situation. The message went something like there is enough ammunition for two days, the Panther tanks are down to one round per day. Fuel is as good as exhausted, and the pocket is about 5 – 6 km long and less than 1 km deep. In it are approx. 25,000 – 30,000 men. Every shell fired by the Russians hits something.

It was just before Easter and Himmler answered us. His message: every man must stand ... already seen through very different situations ... no fatigue and other such slogans. He concluded by wishing every officer, NCO, and enlisted man a Happy Easter. We had never experienced a greater irony. In our situation we did not really expect an order to withdraw, but wishing us a Happy Easter when we were taking heavy casualties every day and had no ammunition and nothing to eat was really too much. Most of us were living on canned sardines in oil, as these could still be had in a navy depot. We still had our kitchen, but it was in a hole, as we no longer had any vehicles. The wounded were evacuated by sea to Hela. One had to have a pass. Many men stood down at the beach near the boats, mainly older sailors, and when the military police asked them for a pass the men showed them their empty ammunition pouches and said that this was their pass.

Acting on his own responsibility, our corps general made an agreement with the navy that we would be taken to Hela on the night of 4-5 April 1945. He declared: 'either I'll be hung from the nearest lamppost in Berlin or I'll be given the Oak Leaves.'"

Under the corps headquarters in Hela, the 4th SS Police Panzer Grenadier Division assumed responsibility for defending the coast in the area around Heisternest. On 9 April 1945 it reported a fighting strength of 3,110 men and a rations strength of

4,871 men. These figures included a Navy rifle battalion temporarily attached to the division and a *Luftwaffe* battalion. There were no more heavy weapons. On account of the division's still "considerable" strength, the OKH ordered it sent to Army Group Vistula.

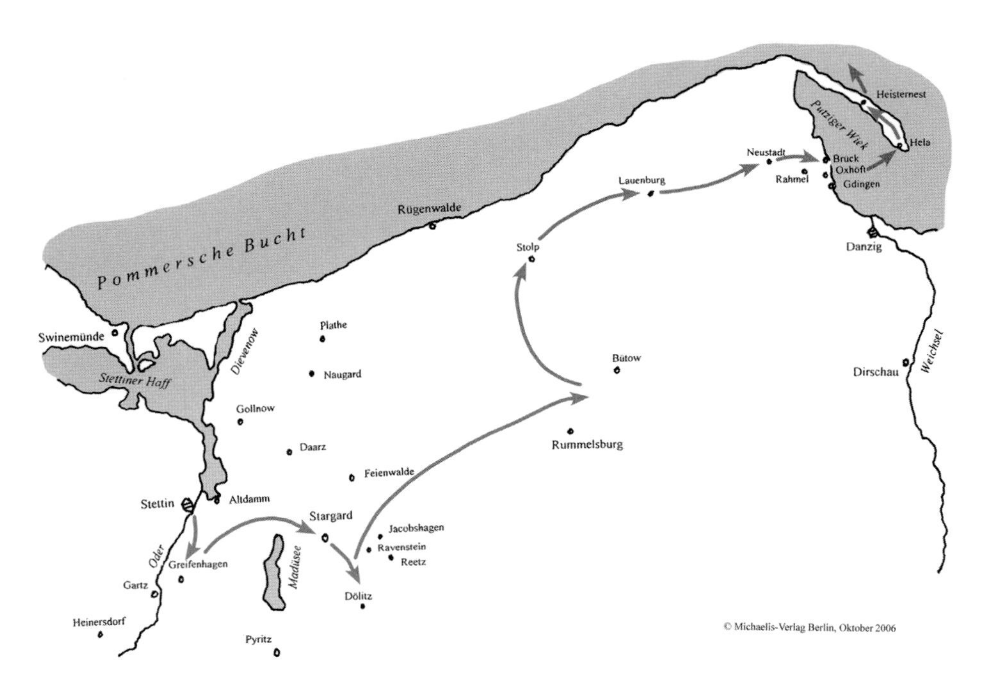

From Stettin to Hela.

SS Battle Group Harzer
(*SS-Kampfgruppe "Harzer"*)

On 13 April 1945 the 4th SS Police Panzer Grenadier Division began boarding ships for transport to Swinemünde. It was being sent to bolster the Oder Front in anticipation of a Soviet offensive. After arriving the next day the units initially assembled near Heringsdorf, and on 18 April were attached to III SS Panzer Corps in the Gramzow area. When Army Group Vistula ordered the corps commander, *SS-Obergruppenführer und General der Waffen-SS* Steiner, to guard the junction between the 3rd Panzer Army and the 9th Army between Liebenwalde and Oderberg, the remnants of the 4th SS Police Panzer Grenadier Division were reorganized and, reinforced by army elements, renamed SS Battle Group Harzer. The battle group was organized as follows:

Headquarters
SS Panzer Grenadier Regiment 7[8]
SS Panzer Grenadier Regiment 8
SS Fusilier Regiment 4
Army Pioneer Battalion 630
(Army) Armored Reconnaissance Battalion 115

SS Panzer Grenadier Regiment 7 marched into the Eberswalde area, the first unit of the battle group to arrive. At the same time the Red Army broke through in the direction of Bernau. The rapid advance by enemy tanks in the direction of Oranienburg led to the order for the rest of SS Battle Group Harzer to be sent immediately into the Oranienburg area instead of to the Eberswalde bridgehead.

Operations in Mecklenburg

And so the battle group went into action divided. As the Soviets had breached the Randow position, SS Panzer Grenadier Regiment 7 was sent to the XXXXVI Panzer Corps in Mecklenburg – Lower Pomerania. After heavy fighting near Prenzlau and Neubrandenburg, what was left of the regiment made its way via Waren and Karow to Hagenow, where it surrendered to American forces. After fierce fighting between Oranienburg and Berlin, on 25 April 1945 SS Battle Group Harzer initially withdrew to the line Kremmen – Behrensbrück – Kuhbrücke forester's house. From there, at the beginning of May 1945 the remains of the former 4th SS Police Panzer Grenadier Division tried to reach the Allied demarcation line near Schwerin – Ludwigslust by way of Kyritz and Perleberg. There they surrendered to the Americans.

Hastily formed as a horse-drawn infantry division in 1939 using young and old police members, in the beginning the division was only usable as an occupation unit. After an exchange of personnel, in Russia the division became a more capable unit that experienced many difficult, but also costly, actions, such as at the Volkhov and outside Leningrad. The order for the formation of a panzer grenadier division could not be completely carried out, on account of the security mission in Greece and shortfalls in weapons and equipment. Decimated to battle group strength, the units took part in the offensive in Pomerania, the fierce fighting in West Prussia, and the final battles in Mecklenburg.

Military Postal Numbers (*Feldpostnummern*):

Division Headquarters	00 386
SS Panzer Grenadier Regiment 7	11 376
I Battalion	12 658
II Battalion	16 982
III Battalion	21 354
SS Panzer Grenadier Regiment 8	29 428
I Battalion	29 948
II Battalion	32 194
III Battalion	34 798
SS Artillery Regiment 4	01 061
I Battalion	02 403
II Battalion	03 780
III Battalion	04 528
IV Battalion	05 145
SS Flak Battalion 4	44 142
SS Panzer Battalion 4	59 381
SS Tank Repair Battalion 4	18 286
SS Signals Battalion 4	20 049

SS Armored Reconnaissance Battalion 4	10 883
SS Anti-Tank Battalion 4	01 475
SS Pioneer Battalion 4	05 429
SS Economic Battalion 4	24 889
SS Medical Battalion 4	59 121
SS Supply Troops 4	56 073

Commanding Officers:

10/39 – 11/40	Generalleutnant Pfeffer-Wildenbruch
11/40 – 08/41	Generalleutnant Mülverstedt
08/41 – 12/41	SS-Brigadeführer Krüger
12/41 – 04/43	SS-Gruppenführer Wünnenberg
04/43 – 06/43	SS-Oberführer Freitag
06/43 – 11/44	SS-Brigadeführer Schmedes
11/44 – 05/45	SS-Standartenführer Harzer

Wearers of the Knight's Cross of the Iron Cross:

11/09/41	SS-Standartenführer und Oberst der Schupo Schulze
15/11/41	SS-Standartenführer und Oberst der Schupo Wünnenberg (Oak Leaves on 23/04/42)
13/12/41	SS-Brigadeführer und Gen.Maj. der Waffen-SS Krüger
11/05/42	Major der Schupo Pannier
15/05/42	SS-Sturmbannführer und Major der Schupo Dörner (Oak Leaves on 16/11/44) (Swords on 01/02/45)
30/09/42	SS-Standartenführer und Oberst der Schupo Giesecke
30/09/42	SS-Sturmbannführer und Major der Schupo Schümers
15/10/42	SS-Hauptsturmführer und Hauptmann der Schupo Dietrich
21/10/42	SS-Unterscharführer und Wachtmeister der Schupo Seitz
28/03/43	SS-Obersturmbannführer u. Oberstltn. Der Schupo Bock
09/12/44	SS-Sturmbannführer und Major der Schupo Prager
23/10/44	SS-Obersturmführer Scherg
31/10/44	SS-Hauptsturmführer Auer
16/11/44	SS-Hauptsturmführer Utgenannt
23/02/45	SS-Sturmbannführer und Major der Schupo Traupe
17/03/45	SS-Sturmbannführer und Major der Schupo Etthöfer
28/03/45	SS-Obersturmbannführer Tappe

The 11th SS Volunteer Panzer Grenadier Division "Nordland"

(*11. SS-Freiwilligen-Panzergrenadier-Division "Nordland"*)

The start of "Operation Barbarossa" on 22 June 1941 triggered discussions between the German Reich and Denmark, which since 1940 had been "*occupied by German troops for its own protection.*" The topic of the discussions was the extent to which Danish volunteers would be permitted to and could participate in the war against the Soviet Union. Denmark rejected the official sending of a Danish troop contingent, but it did authorize its citizens to join an independent unit within the German armed forces.

Himmler planned to integrate the Danish volunteers into Waffen-SS units as "*Germans*"; however, this was opposed by the German envoy in Denmark, Renthe-Fink, who argued that if this was done scarcely any Danes would volunteer. Many Danes welcomed the struggle against the expansionist USSR and had fought on the Finnish side against the Red Army in 1939-40; however, they had little sympathy for the racial-ideological objectives of the Nazis, and in particular the SS.

Nevertheless, after discussions with the Wehrmacht, the *Reichsführer-SS* was able to have the so-called "Germanic volunteers" placed under the command of the *SS-Führungshauptamt* (headquarters of the armed SS). For the reasons previously stated, however, the prefix "SS" was not considered for the unit title.

The formation, dubbed **Volunteer Unit Denmark** (*Freiwilligen-Verband "Dänemark"*), was created on 3 July 1941. Five days later the Danish media announced that active and discharged members of the Danish military could volunteer to take part in the struggle against communism. In addition to Danes, many so-called Northern Schleswigers—ethnic Germans from Denmark—also volunteered.

On 19 July 1941 just under 500 volunteers were discharged in Copenhagen. They were given SS ranks with the prefix "*Freiwilligen*" (volunteer), i.e. *Freiwilligen-Sturmmann* or *Freiwilligen-Oberscharführer.*

Renamed "**Freikorps[9] Danmark**" on 15 August 1941, the unit was transported to Posen-Treskau via Hamburg – Langenhorn. In the weeks that followed the volunteers became split along political lines. While one group was anti-communist but Danish-nationalist, the other group was aggressively National-Socialist. When the *Freikorps*' commander, *Freiwilligen-Obersturmbannführer* Kryssing, had a young Dane arrested for spreading Nazi propaganda and allegedly calling for a mutiny, the SS-FHA stepped in. It replaced Kryssing with *SS-Sturmbannführer* Schalburg, a Danish National-Socialist serving with the SS Division "*Wiking.*"

After about nine months of formation and training, at the end of April 1942 the approximately 800-man-strong *Freikorps Danmark* was transported by air to encircled Demyansk. Attached to the SS Totenkopf Division, on 20 May 1942 the Danes occupied positions on the Lovat River. By the time it was withdrawn from the front on 26 July the *Freikorps* had suffered the following losses in killed and wounded:

Officers	Other Ranks	Total
9	337	346

Moved to Mittau, on 7 September 1942 the Danes were given four weeks annual leave. The propaganda coup desired by the German side failed to materialize—in fact, the exact opposite happened. The provocative behavior exhibited by many members of the *Freikorps* (especially those who were Nazi Party members) led to serious tensions with the Danish population.

The *Freikorps* reassembled at Mittau on 18 October 1942, and on 21 November it was transferred to Bobruisk and attached to the 1st SS Motorized Infantry Brigade (*1. SS-Infanterie-Brigade (mot.)*. With it the Danes took part in the fighting southwest of Velikiye Luki. From an initial strength of about 1,100 men, the following occupied positions along the Ushytsa River to Lake Usho at the end of December:

Officers	Other Ranks	Total
12	630	642

At the end of February 1943, *Freikorps Danmark* took part in "Operation Kugelblitz" (ball lightning) under the overall command of the 3rd Panzer Army. The goal of the operation was the destruction of partisan forces east of the Nevel to Vitebsk rail line. This action, which ended on 3 March 1943, was followed by "Operation Donnerkeil" (thunderbolt), which began on 17 March. Lasting until 2 April, the operation saw German forces comb the area west of the Nevel to Vitebsk railway line around the Obol River. Like the first operation it was not a decisive success. The partisans were able to slip away into the swamps and forests. *Freikorps Danmark* was subsequently ordered to Training Camp "Grafenwöhr," where it was disbanded on 20 May 1943. The remaining approximately 650 Danes formed the cadre of a new formation, the SS Volunteer Panzer Grenadier Regiment "Denmark" (*SS-Freiwilligen-Panzergrenadier-Regiment "Danmark"*).

The Volunteer Legion "Norway"
(*Freiwilligen-Legion "Norwegen"*)

As in Denmark, efforts to recruit for the *SS-Standarte "Nordland"* met with little success in Norway. Not until the war with the Soviet Union broke out was there an increase in recruitment, for then the volunteers included men who, while they had no particular attachment to Germany, were anti-communist.

On 29 June 1941 Hitler authorized the formation of the Volunteer Legion "Norge," and the Reich Commissar for Norway subsequently made a grand announcement on the radio. The volunteers began assembling at Camp Fallingbostel on 1 August 1941 and were sworn in on 3 October. As with the *Freikorps Danmark*, the "SS" prefix was not used in the unit title for political reasons. SS ranks were used, with a prefix indicating their status as legionnaires, i.e. *Legions-Unterscharführer* or *Legions-Untersturmführer*.

By the time the Volunteer Legion "Norwegen" was sent to the Leningrad front at the end of February 1942 1,218 Norwegians had volunteered. On 16 March 1942 the battalion reached L Army Corps' area and was attached to Police Battle Group Jeckeln at the junction with the XXVI Army Corps north of Krasnoye Selo. The Red Army launched a major attack on 24 August 1942, and the Volunteer Legion "Norwegen" became involved in fierce and costly defensive fighting.

At the end of November 1942 the Norwegians were placed under the command of the 2nd SS Motorized Infantry Brigade. On 31 December 1942 the Norwegian unit reported a strength of:

Officers	Other Ranks	Total
20	678	698

After heavy winter fighting, in March 1943 the battalion was pulled out of action and moved to Mitau. From there the approximately 600 remaining volunteers were sent to Training Camp Grafenwöhr. After the legion was disbanded, on 20 May 1943 about half of these formed the cadre for a new unit, the SS Volunteer Panzer Grenadier Regiment "Norge." The other half returned to Norway or were transferred to other units. John Sandstad was a member of the Volunteer Legion "Norge":

"I was born in Hunan, China, on 5 February 1925, the son of a Norwegian missionary. In 1932 my father had to leave China for health reasons and bought a small farm in Norway. Following the Soviet attack on Finland in 1939-40, I came to perceive an imminent threat to the northern countries and Western Europe, and in 1942 I volunteered to serve in the Volunteer Legion 'Norwegen.' Not long before, two friends of mine had signed up and been called up in a week or two. I wasn't called up for six weeks, which angered me greatly, because I had wanted the three of us to be together.

I was examined at the Waffen-SS replacement detachment in Oslo and afterwards travelled to Holmestrand, southwest of Oslo, with two boys from eastern Norway. My service began on 10 July 1942 at the training camp of the legion's replacement unit. There was a half-trained rifle company there and a heavy company was in the process of being formed. I was assigned to the latter. At the beginning there were only 20 to 30 men, but more arrived each week. The rifle company was transferred to the Leningrad front at the beginning of August 1942.

Approximately 80 men strong, our heavy company went to Mitau, in Latvia, at the beginning of October 1942. There we were quartered in a former hunting lodge or estate about two kilometers outside the city. We also received a few new people there. The training was hard, the instructors German, as was the language of command. In general we and our German instructors got along well. Only our platoon leader, an SS-Oberscharführer, was rather bad-tempered, as we came to find out.

We became ready for action in January 1943, but a short time later the Volunteer Legion "Norwegen" was pulled out of action. Our company was thus a replacement unit without a frontline unit. While some of us were sent for special training to Dresden (pioneers) and Holland (infantry gun), I and several others were selected to be assistant instructors for new Norwegian volunteers.

On 16 March 1943 the Reichsführer-SS inspected us at Mitau. After the inspection he gave a speech in which he spoke about the future division of the 'Norge' Regiment. We were able to ask him questions in a casual atmosphere. Someone asked him about the status of the so-called 'Jewish question.' Himmler told us that negotiations were under way between Germany and France concerning the establishment of a Jewish state in Madagascar. Although he was no soldier—slightly built and a wearer of glasses—at the time I found him quite likeable. At the end of March 1943 we were moved to Grafenwöhr/Auerbach."

The 11th SS Volunteer Panzer Grenadier Division "Nordland"
(*11. SS-Freiwilligen-Panzergrenadier-Division "Nordland"*)

After the grievous losses suffered at Stalingrad and in Africa, efforts were made to establish new units. In May 1943 an agreement was reached between Germany and Rumania concerning the recruitment of ethnic Germans for the Waffen-SS.

This new and extensive potential made it possible to immediately form a new division. Planning for the new unit proceeded in parallel with the German-Rumanian talks, and even before the agreement was ratified on 22 March 1943 Hitler ordered the formation of a division for the III (Germanic) SS Panzer Corps.

For pure propaganda reasons Himmler gave the unit the name "*Nordland*," in order to create the impression that the division was made up of volunteers from northern Europe. Even though the *Freikorps "Danmark,"* the Volunteer Legion "*Norwegen*," and the SS Panzer Grenadier Regiment "*Nordland*" of the SS Panzer Grenadier Division "*Wiking*" were all incorporated into the new division, northern Europeans only made up about 15% of its personnel.[10]

At the beginning of 1943 the Waffen-SS planned the formation of the

11th Latvian SS Volunteer Division
12th Lithuanian SS Volunteer Division
13th Croatian SS Volunteer Division

and the "*Nordland*" Division was initially given the title:

14th (Germanic) SS Panzer Grenadier Division "Nordland"
(*14. (germanische) SS-Panzergrenadier-Division "Nordland"*)

Formation of the three new divisions was delayed, however, and on 10 April 1943 the unit's title was changed to:

SS Panzer Grenadier Division 11 (Germanic)
(*SS-Panzer Grenadier Division 11 (germanisch)*)

Finally, the division received its ultimate title:

11th SS Volunteer Panzer Grenadier Division "Nordland"
(*11. SS-Freiwilligen-Panzergrenadier-Division "Nordland"*)

Training in Croatia, 1943.

The division was organized as follows:

Division Headquarters
SS Panzer Grenadier Regiment "*Norge*" (I – III Battalion)
SS Panzer Grenadier Regiment "*Danmark*" (I – III Battalion[11])
SS Panzer Artillery Regiment (I – III Battalion)
SS Panzer Flak Regiment
SS Panzer Battalion "*Hermann von Salza*"[12] (Panzer IV)
SS Assault Gun Battalion
SS Armored Signals Battalion
SS Armored Reconnaissance Battalion
SS Anti-Tank Battalion
SS Armored Pioneer Battalion
SS Economic Battalion
SS Field Replacement Battalion
SS Medical Battalion

After the change of power in Italy and the announcement that the country's occupation forces in the Balkans would be withdrawn, on 2 August 1943 Hitler ordered the transfer of large elements of the Replacement Army to fill the predictable vacuum. One unit affected by this move was the 11th SS Volunteer Panzer Grenadier Division, part of the future III (Germanic) SS Panzer Corps. On 20 August 1943 the division began moving from Training Camp Grafenwöhr to the area south of Sisak (approx. 50 km southeast of Agram).

In addition to the formation and training of units—many ethnic Germans did not arrive until the summer of 1943—the division took part in sometimes difficult operations against communist partisans and the disarming of Italian units in the Sambor and Karlovac area. The Norwegian volunteer Ingebret Lilleborge remembered the formation period:

"*We had actually volunteered to serve in Finland. Some of us were therefore very dissatisfied with the transfer to Croatia. Some were even discharged from the SS Panzer Grenadier Regiment 'Norge' and sent to Finland. New replacements arrived from Norway and this calmed the agitated German command.*

I had just returned from a course in Kostajnica and was summoned by my battalion commander: 'You're staying here to become a proper soldier!'

Relations with the population were good in Croatia. Behind our positions life was quite peaceful and normal—after all, we were shielding them from the terror of the partisans."

Swearing-in ceremony in Hrastelnica, 4/10/43.

In October 1943 the regiments and divisions of the Waffen-SS were renumbered. The two panzer grenadier regiments were given the numbers 23 and 24.

At the end of November 1943 Hitler ordered the III SS Panzer Corps to the Eastern Front. In December the 11th SS Volunteer Panzer Grenadier Division "*Nordland*" arrived in the Kirova area at the Oranienbaum pocket (18th Army). Its strength at that time was:

	Officers	NCOs	Enlisted Men	Total
Authorized	341	1,975	10,146	12,462
	2.7%	15.8%	81.5%	100%
Actual	618	2,877	10,704	14,199
	4.3%	20.3%	75.4%	100%

Equipment strength stood at:

	Authorized	Actual
Assault guns	31	30
Armored troop carriers	167	141

In December 1943 the division commander, *SS-Brigadeführer* von Scholz, assessed his unit as follows:

"*The state of training of the division, which was formed in June-July 1943, largely from recruits, was such that, after a brief period of basic training, the division was transferred to the training camp in Croatia for further training and to guard against partisans. The nature of the mission only allowed unit training up to platoon strength. The assigned security missions favored training in defense. Since the move to the Oranienbaum pocket unit training has been carried out by the reserve battalions, which are rotated at three-week intervals. Unit training should therefore be completed in 18 weeks (end of April 1944).*

Morale of the troops outstanding.

The inadequate allotments of fuel are causing difficulties, as motorized unit training is impossible. Personnel shortages, felt especially by the artillery, must be addressed.

The division is capable of limited offensive missions and is conditionally suitable for defensive missions."

After launching its winter offensive, on 14 January 1944 the Red Army broke into the positions of the 9th and 10th Field Divisions (*Luftwaffe*) deployed at the east end of the pocket, and elements of the 11th SS Volunteer Panzer Grenadier Division were immediately dispatched to restore the situation. The Red Army enjoyed a massive superiority, and its attacks led to the collapse of the German front. In extremely difficult and costly fighting the units tried to reach Narva, about 75 kilometers to the southwest. Franz Bereznyak, an ethnic German SS volunteer, summed up the fighting:

"*We reached Narva on 30 January 1944. Who would have thought that our proud unit would look like this after just two weeks?*"

Heinz Twesmann, then an *SS-Hauptscharführer*, recalled:

"*Our company commander von Bargen was leading us toward Gostilitsy in the darkness, when suddenly we heard machine-gun fire from a wood about 100 meters to our left and Luftwaffe soldiers came running out. Von Bargen immediately ordered me to reconnoiter toward the wood. Cautiously we felt our way forward, until at the north end of the wood we saw six T-34 tanks before us. They failed to see us, however, and we went into position about 50 meters in front of them. At daybreak the tank on the far left began firing on the Rendemann Company. Our company took little fire, but von Bargen was wounded and I assumed command. I called battalion for anti-tank guns, but they were shot to pieces before they could fire even a single shot.*

Then the order came to disengage from the enemy and assemble south of the wood. A battery of assault guns arrived, with which I was supposed to attack across a forest clearing. I climbed onto the assault gun commanded by SS-Hauptsturmführer Ellersiek, who was peering through a scissors telescope, observing the terrain before us.

Suddenly our vehicle was hit and we were thrown to the ground. We began to run and made it to our old positions at the edge of the forest. SS-Hauptsturmführer Ellersiek was wounded but was later recovered. The T-34s in front of us now began firing at us, whereupon I ordered my men to shoot at the vision slits in the tank cupolas with their carbines. In any case we were able to fall back.

We went into position in a depression due south of the forest's edge and immediately came under directed mortar fire and a salvo from Stalin Organs. Luckily we suffered no casualties. I suspected [there was] a Soviet forward observer in the trees and had all of the company's machine-guns open fire on the treetops. Soon afterwards we saw a Red Army soldier fall to the ground. From then on it was quiet, and we were able to fall back to the battalion command post. It was in an earth bunker next to a Luftwaffe 88-mm flak battery. No sooner had we gone into position when we began taking heavy artillery fire from big guns. One of the guns

The commander of SS Armored Reconnaissance Battalion 11, SS-Sturmbannführer Saalbach, and members of the battalion (above).

Heinz Genzow

Positions on the Gulf of Finland in February 1944.

Living bunker.

Red Army soldiers killed in front of the German lines, February 1944.

next to me took a direct hit, and there were dead and wounded. I was seriously wounded, with a fragment in my lung. That was 17 January 1944."

On 4 February 1944 the newly-created Army Detachment Narva (*Armee-Abteilung "Narwa"*) took command over III SS Panzer Corps and XXVI Army Corps. After light fighting the situation temporarily stabilized along the Narva. *SS-Obersturmführer* Lorenz recalled:

"*All of a sudden Tiger tanks appeared. I remember thinking, my God, they're going to crush us, for a tank had once ripped the entire box off the side of my four-wheeled armored car because the road was too narrow. They came from the Russian lines, God knows from where, and some were towing other tanks. But these Tigers prevented the Russian tanks from finishing us off.*"

In mid-February 1944 the 11th SS Volunteer Panzer Grenadier Division "*Nordland*" received replacements and was able to replenish its decimated units. The unit submitted the following strength report at the end of February 1944:

	Officers	NCOs	Enlisted Men	Total
Actual	102	1,785	9,247	11,134
	0.9%	16%	83.1%	100%
Authorized	564	3,028	11,545	15,137
	3.7%	20%	76.3%	100%

Despite the arrival of more than 1,000 replacements, the unit's actual strength had fallen by about 1,500 men compared to what it had been at the turn of the year 1943-44. More than 2,500 men had been killed, wounded, or captured in two weeks of fighting! The two first battalions of the panzer grenadier regiments were disbanded. Their remaining personnel were divided among the second and third battalions and transferred to SS Training Camp "Hammerstein" in Pomerania for reorganization.[13] The regiments were thus left with two battalions each, the same as Wehrmacht panzer grenadier regiments.

While the positions along the Narva were held—SS Panzer Grenadier Regiment 24 "*Danmark*" was right in the Narva bridgehead—elements of the division were moved southwest to stop enemy forces that had broken through toward Vaivara and Auvere. The Red Army's advance was halted southwest of Narva, but the fighting was ferocious and costly. The front in Army Detachment Narva's area quieted down until summer 1944.

Awarding of the Knight's Cross of the Iron Cross in Hungerburg (Narva-Jöesuu) on 19/3/44. From left to right: SS-Obersturmführer Langendorf, SS-Hauptsturmführer Seebach, and SS-Hauptsturmführer Krügel.

An SS-Unterscharführer is awarded the Iron Cross, Second Class.

On 30 June 1944 the 11th SS Volunteer Panzer Grenadier Division "*Nordland*" reported its strength as follows:

	Officers	NCOs	Enlisted Men	Total
Actual	355	1,877	8,788	11,020
	3.2%	17%	79.8%	100%
Authorized	612	2,877	10,704	14,193
	4.3%	20.3%	75.4%	100%

The division's personnel strength had dropped again since the end of February 1944. In the course of their summer offensive, on 10 July 1944 the Red Army achieved a breakthrough at the junction between Army Groups North and Center in the area north of Vilnius and south of Dvinsk. SS Armored Reconnaissance Battalion 11, which was equipped with a large number of armored troop carriers, was alerted, and on 11 July 1944 was transferred to the hotspot southwest of Dvinsk, about 500 kilometers away.

While the rest of the division was heavily engaged on the Narva front, SS Armored Reconnaissance Battalion 11 reached Lithuania via Latvia. Moving constantly, it simulated larger German formations in an area almost devoid of troops between the two army groups. The battalion fought many engagements with elements of the Red Army advancing on Riga. Hans Stemper remembered the actions in Lithuania:

"*During the trip—at least 800 km—we enjoyed the beautiful countryside in magnificent weather. When we drove through Jakobstadt—on wheels again—we found the city seemingly uninhabited. Our vehicles assembled at an intersection of the main street. While we waited we discovered that the shops had been looted. The doors were open and goods lay everywhere, even on the sidewalk.*

Our mission south of the Dvina was to stop the Soviet spearhead advancing toward Mitau – Tukkum, if possible destroy it, and cover Army Group North's southern flank. Our route now took us south. We crossed the Lithuanian border near Akniste. Not until we were south of Urli, about 130 km southwest of Dvinsk, did the battalion turn west, then north at Panevezys. There were frequent skirmishes, and we were constantly in danger of being outflanked.

On the evening of 26 July 1944 we halted at a lake near Satkunai. The only building was a school. Judging from the facilities, library, gymnasium, etc., it must have been an exclusive boarding school. There was also a well-stocked larder. Beneath the ceiling hung hams, sides of bacon and sausage—all hard to believe. Everything was clean and empty, as if the children had just left school. We never

75-mm Type 18 infantry gun.

Assault guns roll toward the front. Sooküla near Põlva, April 1944.

The division commander, SS-Brigadeführer Fritz von Scholz, on the Narva front.

Nationality badges for the volunteers from Norway (left) and Denmark (right).

The cuff title of the 11th SS Volunteer Panzer Grenadier Division "Nordland."

thought of looting the larder, probably because most of us were receiving regular parcels from home, meaning from the Banat and Transylvania. Of course we also scouted the immediate surrounding area, discovering about 300 members of the Todt Organization in a hollow. They were sitting or lying around apathetically and had already resigned themselves to being captured. Of course they were extremely happy when we appeared.

From south of the lake we could hear the sounds of battle—as a precaution the commander's vehicle was driven into cover behind the school. We were warned and placed on alert... All of a sudden we learned that the enemy had broken through to the south. There was only one avenue left to us, north, over a hill. On 27 July 1944 the entire SS Armored Reconnaissance Battalion 11 formed up, as if it were on the training grounds.

Suddenly SS-Obersturmführer Langendorf shouted: anti-tank guns in position, fall out at once! He and the leader of the company headquarters squad sprinted left into the ditch at the side of the road, Borger and I to the right. Bent low, we moved forward, firing as we went. Only Obersturmführer walked upright, as always, observing the cornfield. We had already come to believe that he was invincible.

Kneeling down to reload, I looked to my left to see if the battalion had closed up. Then something hit me in the face and suddenly I saw brightly-colored circles, rushing toward me ever faster and larger. A sniper had shot me. I thought—now they've got you—what will mother say—no, you mustn't die now! I turned around and showed Borger my packet dressing. Because of my wound I couldn't speak. Back at the armored troop carrier the driver pulled me inside and applied an emergency dressing. He drove me back to the first-aid vehicle in reverse. The medic, a fellow countryman—our birthplaces in Transylvania were just 15 kilometers apart—talked to me soothingly. He replaced the dressing and, after I made myself understood with gestures, gave me an injection. I passed out, but then I was awakened by a shout: breakthrough successful, we've done it!"

Although the German troops succeeded in stopping the Soviet advance in Latvia-Lithuania and reestablished a continuous front, on 25 July 1944 Red Army pressure on the Narva front resulted in a withdrawal to the so-called Tannenberg Position between Vaivara and Auvere. In the days that followed there was bitter, costly fighting for possession of the three strategically-important hills[14] in the position.

On 31 July 1944 SS Armored Reconnaissance Battalion 11 was pulled out of action in Latvia and ordered back to the division in the Tannenberg Position near Vaivara. There was fierce fighting for possession of the individual hills until the beginning of August. Unable to overcome the Tannenberg Position, the Soviet

Estonia, summer 1944.

command shifted the focus of the attack to the south. On 1 August 1944 the OKW reported:

"Yesterday, the enemy did not continue his massive assault in the Narva Isthmus due to heavy losses. Weaker attacks failed. The III (Germanic) SS Panzer Corps commanded by SS-Obergruppenführer und General der Waffen-SS Steiner, with the Germanic SS Volunteer Divisions "Nordland" and "Nederland," the 20th Estonian Volunteer Division, and the 11th East Prussian Infantry Division played a major role in the defeat of the Russian offensive in recent days, along with units of the Kriegsmarine fighting on the land front and army artillery and rocket units."

On 10 August 1944 the Red Army broke through the German front south of Lake Pskov and reached the area east of Vöru. A drive northwards was supposed to encircle the German troops in the Tannenberg Position. Army Detachment "Narva" subsequently received orders to halt the enemy advance between Lake Wirz and Lake Peipus. Four days later Battle Group Wagner, a division-strength formation made up of army troops, elements of the 4th SS Volunteer Panzer Grenadier Division "*Nederland*," Estonian SS volunteers, and SS Armored Reconnaissance Battalion 11 was able to establish a main line of resistance along the Embachs after fierce fighting. After the Red Army succeeded in establishing a bridgehead near Tartu (Dorpat) the fighting temporarily died down.

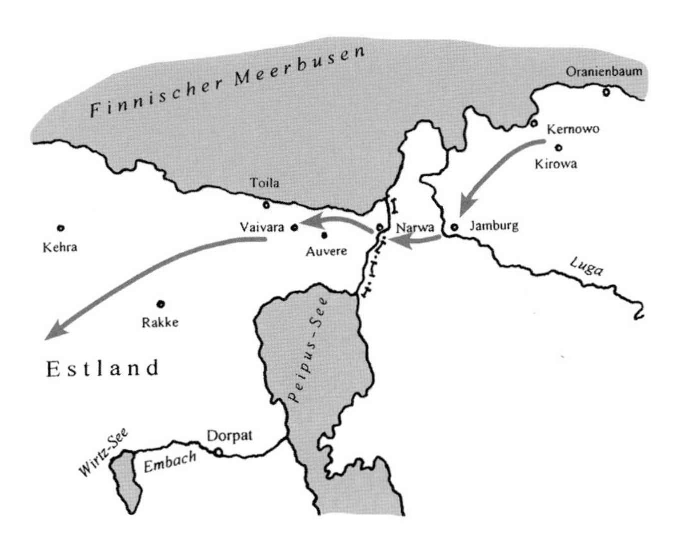

Sanatorium Hill in the Blue Mountains (Estonia).

In order to avoid Army Detachment Narva from potentially being cut off, however, in mid-September 1944 Hitler authorized the evacuation of Estonia. The 11th SS Volunteer Panzer Grenadier Division "*Nordland*" withdrew from the Tannenberg Position and on 21 September 1944 crossed the Estonian-Latvian border near Ainazi. The division's strength at that time was:

	Officers	NCOs	Enlisted Men	Total
Actual	328	1,818	8,334	10,480
	3.1%	17.3%	79.6%	100%
Authorized	558	3,391	12,612	16,561
	3.4%	20.5%	76.1%	100%

Ordered into the Baldone area, the division immediately attacked to relieve the beleaguered capital. The result was fierce fighting with terrible losses on both sides. The Red Army, however, halted its advance on Riga and switched the emphasis of its attack to the Dobele area. The 11th SS Volunteer Panzer Grenadier Division was again pulled out of its positions and moved into the new attack zone by way of Jaunpils. After further heavy fighting the unit was moved again; after enemy troops reached the Baltic Sea near Polangen the SS units were ordered to assemble in the Priekule area and prepare for an attack towards the south.

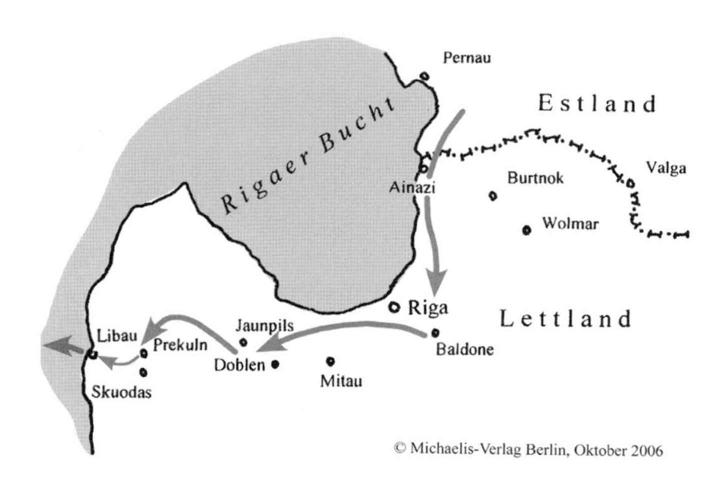

This plan remained just that, for 16 October 1944 saw the beginning of the so-called First Battle of Courland. The Soviet offensive struck the 11th SS Volunteer Panzer Grenadier Division with full force. With great effort the men were able to hold their positions. The fighting died down temporarily until 27 October 1944 when, after a tremendous bombardment, the Red Army made another attempt to break into the German positions. The Second Battle of Courland had begun. The offensive ended at the beginning of November 1944. The enemy then shifted his focus once again. Edi Janke, then an *SS-Unterscharführer*, recalled:

"*One day in December 1944 a fierce snowstorm began. I warned the entire platoon that the Russians might come in white camouflage clothing and urged everyone to remain vigilant. In the evening I lay down to rest and was subsequently awakened by the machine-gunner's screams: Unterscharführer, it's the Russians! I grabbed the flare pistol. When I fired a white illumination flare about 30 Red Army soldiers immediately froze. At once I fired a second flare. The machine-gunner on the right swung his weapon around and fired an entire belt into the group. When I fired a third flare we saw the Russians running away. We subsequently discovered one wounded and nine dead Russians. They had cut their way through our barbed wire and had already removed a mine from in front of our positions. I reported what had happened, and the company commander decorated the vigilant machine-gunner with his own Iron Cross, First Class. The name of the gunner: SS-Sturmmann Konrad.*"

On 23 January 1945, during the Fourth Battle of Courland, the focal point of Priekule was again attacked. The enemy succeeded in breaking into the positions several times and a number of units were overrun. The division received support from the 14th Panzer Division and was able to hold the area. It was the 11th SS Volunteer Panzer Grenadier Division's final battle in Courland. Transport by sea to Stettin began at the end of January 1945.

Army Group Vistula was being formed in Pomerania to hold the area between Stettin and Danzig. Also newly formed was the attached 11th Army, which held the area from the mouth of the Finow Canal into the Oder to the line Jastrow – Ratzebuhr – Neustettin with little more than battle groups and elements of the replacement army. Despite the losses suffered in the recent costly fighting, III SS Panzer Corps, just transported out of Courland, was the most capable component of the 11th Army, and on 5 February 1945 it took command in the Freienwalde – Neu-Wedell area.

There was no possibility of reorganizing the units in peace, but in preparation for the imminent Pomerania offensive, on 12 February 1945 the 11th SS Volunteer Panzer Grenadier Division "*Nordland*" began a period of rest and reorganization. SS Panzer Battalion 1 was brought up to regiment strength through the attachment

SS-Untersturmführer Pehrsson, the Swedish commander of 3rd Company, SS Armored Reconnaissance Battalion 11. "He was not the type to sacrifice his men unconditionally. He always considered carefully and was an example, even in battle, always a man!"

The Swedish war correspondent SS-Untersturmführer Krueger.

Hans-Karl Erlewein was transferred from the SS Division "Wiking" to the SS Panzer Grenadier Division "Nordland."

of the 503rd SS Heavy Panzer Battalion. SS Panzer Regiment 11's equipment strength was:

I Battalion:
30 assault guns
30 Panzer V Panther
II Battalion:
39 Panzer VI Tiger II

Interestingly, the unit was not renamed 11th SS Panzer Division "*Nordland*."

Just two days after rest and reorganization began, the approximately 6,000-man-strong division was placed on alert, and on 15 February 1945 it moved into the assembly area for "Operation Solstice" (*Unternehmen Sonnenwende*). The offensive in Pomerania began the next day. The 11th SS Volunteer Panzer Grenadier Division attacked from the Reetz area toward Arnswalde (approx. 12 km southwest of Reetz), which had been encircled by the Red Army.

After the garrison and civilians had been evacuated the troops abandoned the town and withdrew to its start positions, all the while engaged in fierce fighting with the steadily reinforced Red Army.

On 1 March 1945 the Soviets launched an offensive against the German units in Pomerania. The entire German front collapsed within an hour. Heinz Genzow, then an *SS-Sturmmann*, recalled:

"*The town of Jakobshagen was evacuated and a new blocking line was supposed to be established. Our path also took us through Vossberg, a small village on a rail line. While our commanding officer and the company commanders gathered for a situation briefing, some of us went looking for something to eat in the abandoned houses or tried to grab a little sleep. It was like a Sunday afternoon in the village—the first green was appearing on the trees and everything was very peaceful. My comrades and I searched a farmhouse, and as we were foraging we suddenly came upon two Russians. They immediately fired their submachine-guns and hit one comrade in the belly. This gave us an unbelievable shock, and at first we were completely unable to react. This put an end to the quiet in the village, and there was firing from one end to the other. The Russians were shooting from all sides; only the railway embankment and the autobahn underpass were still clear. But several of the enemy's T 34s were already approaching. But now and then miracles happen, and so two Tiger tanks suddenly appeared. I don't know where they came from, but one took up position behind a barn and the other behind a haystack. They immediately opened fire, and I believe almost every shot was a hit. Several enemy tanks were knocked out in no time and the enemy withdrew.*

Sd.Kfz 250/1 light armored car.

In our armored troop carriers, we broke through the autobahn underpass and reached another village. There we saw some Russians in the process of setting up an anti-tank gun. The halftrack to our left turned straight towards it and simply ran it over. The Russians fled in all directions, but there were always more Red Army men. At the exit from the village we encountered a refugee column and took some of the civilians with us. While driving through I took a shell fragment in the thigh and was ultimately taken by ambulance to the aid station."

Constantly on the move, on 7-8 March 1945 the 11th SS Volunteer Panzer Grenadier Division reached the area around Altdamm, which was envisaged as an Oder bridgehead, east of Stettin. On 5 March 1945 Himmler had sent a letter to the commanding general of III (Germanic) SS Panzer Corps, *Generalleutnant* Unrein, and the commander of the 11th SS Volunteer Panzer Grenadier Division "*Nordland*," *SS-Brigadeführer* Ziegler, concerning the action at Arnswalde:

"*I hear that Nordland did not attack and fight well. Crack down hard. I expect Nordland to do its duty as before.*"

He did the division an injustice. The morale of the men was good despite inadequacies in equipment and armaments. On 26 March 1945 the Commander-in-Chief of the 3rd Panzer Army, *General* von Manteuffel, himself made the following assessment of the division:

"*Seriously under strength in panzer grenadiers, under strength in officers and NCOs, request should be made to the RF-SS for priority allocation of replacement personnel. Good divisional artillery, outstanding reconnaissance battalion (78 armored cars). Tactically attached to the division at present: half of SS Heavy Panzer Battalion 503 with 6 serviceable Panzer VI and 7 under repair. Division has outstanding fighting spirit, and after it receives replacement personnel and equipment it will be capable of any offensive mission.*"

Hitler intended to take the offensive in the direction of Danzig from the bridgehead (distance of about 300 km) at a later date, and the units deployed there were bled by Soviet artillery and aerial bombardment. Not until 19 March 1945 did Hitler authorize the abandonment of the completely senseless bridgehead across the Oder. The remaining elements of the 11th SS Volunteer Panzer Grenadier Division withdrew into the area southwest of Stettin during the night of 2 March 1945.

There the unit was supposed to be reorganized as a so-called Panzer Division 45, as per a decree by the Inspector-General of Armored Forces. It remained a theoretical order. The units were, however, brought up to strength with replacements from the air force, navy, and Waffen-SS. Among them were a few members of the "British Free Corps."[15]

Total casualties in the period from 1 September 1943 to the end of March 1945 were:

	Officers	Other Ranks	Total
Killed	102	2,837	2,939
Wounded	267	10,005	10,272
Missing	23	1,255	1,278

With a total number of 14,489 casualties the division had suffered losses equivalent to its entire personnel complement.

On 27 March 1945 the unit moved with the III SS Panzer Corps into the area north of Angermünde. Nine days later the 11th SS Volunteer Panzer Grenadier Division "*Nordland*" submitted the following serviceability report:

2 Panzer V Panther
26 Sturmgeschutz III
13 Panzer VI Tiger
10 anti-aircraft tanks
11 75-mm anti-tank guns (towed/self-propelled)

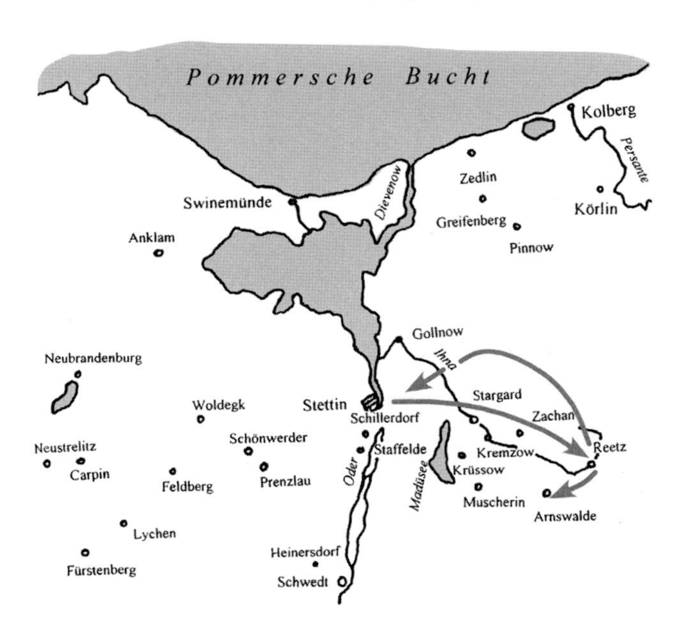

75-mm Pak 40 anti-tank gun.

On 16 April 1945 the Soviets launched "Operation Berlin." In response, the 11th SS Volunteer Panzer Grenadier Division "*Nordland*" was ordered into the area south of Frankfurt/Oder. The move temporarily ended in the Strausberg area because of a shortage of vehicles and fuel. There the unit was attached to LVI Panzer Corps.

Two days later the division occupied defensive positions in the area in front of Strausberg. The only tank battle at the gates of Berlin was fought the next day. Near Prötzel, SS Panzer Regiment 11 destroyed about 100 enemy tanks. In recognition of this success, on 23 April 1945 the regimental commander, *SS-Obersturmbannführer* Kausch, was awarded the Knight's Cross with Oak Leaves (846th recipient).

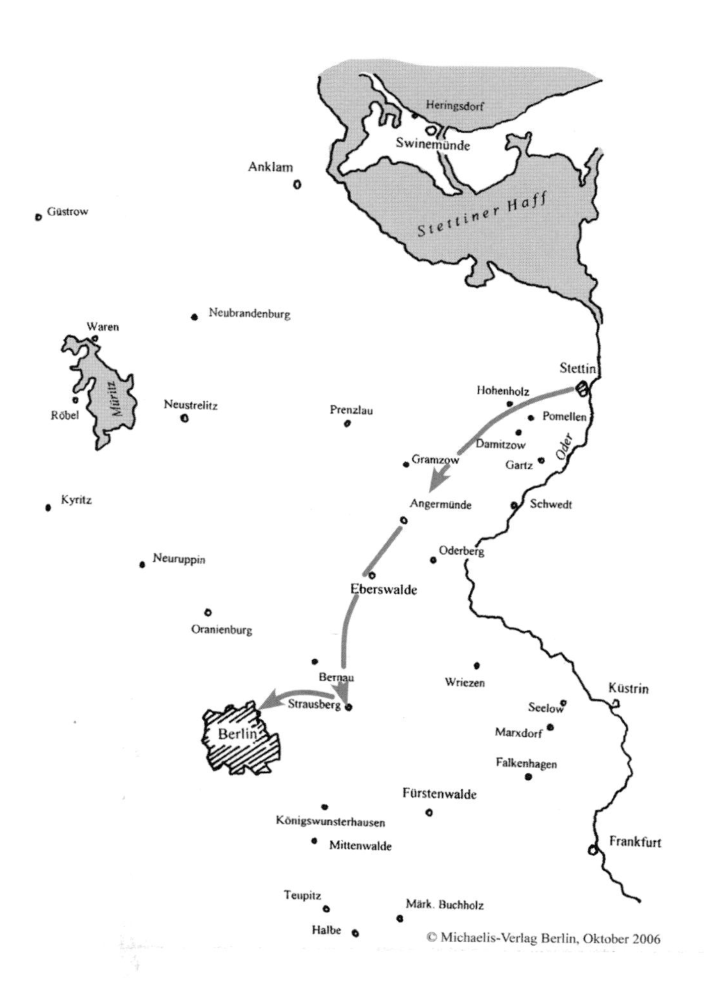

Exhausted, the soldiers withdrew through Mahlsdorf toward Berlin. As part of LVI Panzer Corps, the 11th SS Volunteer Panzer Grenadier Division was supposed to help defend the Reich capital. By then its fighting strength was about 1,500 men[16]—roughly equivalent to a regiment.

Constantly under fire from Soviet tanks fighting their way into the center of the city, on 23 April 1945 the units crossed the Spree in Treptow and took up position in a new defensive front along the streetcar line between Treptower Park and Tempelhof. Two days later the commander of the 11th SS Volunteer Panzer Grenadier Division, *SS-Brigadeführer* Ziegler, was placed under arrest for insubordination. He was replaced by *SS-Brigadeführer* Dr. Krukenberg, who brought with him 90 French SS volunteers. The then *SS-Unterscharführer* Burgkart recalled:

"On the morning of 25 April 1945 we were standing in front of the stairs leading to the division command post—a building in the Hasenheide. Suddenly someone spoke to us: Where is the 'Nordland' command post? I turned around and said: down there in the cellar. And as I said this I saw the silver-grey trim on the coat and realized that it was an SS-Brigadeführer. He was standing at the corner of the house, slightly behind us to the right. Then, without saying a word, he passed by us followed by a few SS men carrying submachine-guns.

When I looked around, I saw that several trucks carrying SS people had pulled up, unnoticed by us. The men had got down from the trucks and had formed a skirmish line, sealing off the entire street. Everyone was stopped. A short time later SS-Brigadeführer Ziegler came up the stairs with his driver, SS-Hauptsturmführer Emmert, and his batman. Ziegler came right up to me and said: Burgkart, take your mess kit out of the car and if you have any other personal things, those too. I asked: Why then, Brigadeführer? His answer: Take your things out, I have to go!

Only then did we learn that the second SS-Brigadeführer was Dr. Krukenberg of the 33rd SS Waffen Grenadier Division 'Charlemagne' and that the SS men with him were French.

By then Emmert had climbed into the Schwimmwagen and started the motor. I took my mess kit, my assault rifle, and my heavy jacket out of the vehicle. Then Ziegler and his batman got in. He stood in the front seat of the vehicle, saluted us, and said to the two SS-Sturmbannführer standing next to me: All the best gentlemen!

When he had gone, SS-Sturmbannführer Vollmar looked at me and said: What's happening now? And then again: What's happening now? I left to take care of a few things, and when I tried to return to the command post I was stopped by one of the Frenchmen: No, back, no, back!

Who wants to go back?, I shouted. At that moment, with no one able to understand the other very well, two or three armored troop carriers carrying wounded SS men pulled up. They had come from the area south of the Hasenheide. The crews were looking for a hospital for their groaning, crying, and screaming comrades. Halt, stop, stop, the Frenchmen again shouted. The drivers ignored them and drove on. Suddenly one of the Frenchmen raised his assault rifle and fired at the first armored troop carrier. The co-driver manning the MG 42 reacted by simply firing into the group of French SS men. I saw three or four Frenchmen fall to the ground and cry out. OK, I thought to myself, get out of here now, as quickly as possible. I drove to the new command post, the State Opera 'Unter den Linden,' with what was left of the operations officer's staff. There we were divided into squads and deployed at the Spittelmarkt."

With no continuous front, the men fought their way to the Air Ministry and then to the Weidendammer Bridge—sometimes contesting individual floors of buildings. There, on the night of 1 May 1945, mixed military units and civilians tried to break through the ring of Soviet positions. In several waves, which included the last German tanks, they tried to force their way across the bridge and through the adjoining blocks of houses. All attempts collapsed under heavy enemy fire.

What was left of the 11th SS Volunteer Panzer Grenadier Division "*Nordland*" was captured by the Soviets. Thus ended the story of a remarkable unit. Contrary to its title "*Nordland*," the division was largely made up of ethnic Germans from Rumania. Even in 1943 there had been problems supplying the unit with the weapons, equipment, and vehicles needed for operational training. During its very first deployment at the front the division became involved in a major battle, helping fight off the Soviet winter offensive at the Oranienbaum pocket. Despite heavy losses, morale remained high thanks to the division command. During its brief period of operations the 11th SS Volunteer Panzer Grenadier Division "*Nordland*" was awarded more Knight's Crosses than any other unit of the Waffen-SS.

Military Postal Numbers

Division Headquarters	33 316
SS Supply Unit 11	38 826
SS Panzer Grenadier Regiment 23 "Norge"	41 891
I Battalion	32 298
II Battalion	42 264
III Battalion	32 878
SS Panzer Grenadier Regiment 24 "Danmark"	35 408
I Battalion	40 670
II Battalion	37 826
III Battalion	34 531

SS Volunteer Artillery Regiment 11	42 973
I Battalion	35 179
II Battalion	32 895
III Battalion	34 885
SS Panzer Battalion 11 "Hermann von Salza"	32 192
SS Tank Repair Battalion 11	36 479
SS Anti-Tank Battalion 11	32 356
SS Armored Reconnaissance Battalion 11	33 756
SS Economic Battalion 11	39 384
SS Pioneer Battalion 11	38 749
SS Flak Battalion 11	43 111
SS Signals Battalion 11	48 843
SS Field Replacement Battalion 11	59 858

Commanding Officers

05/43 – 07/44	SS-Brigadeführer von Scholz
07/44 – 04/45	SS-Brigadeführer Ziegler
04/45 – 05/45	SS-Brigadeführer Dr. Krukenberg

Wearers of the Knight's Cross of the Iron Cross

18/01/42	SS-Oberführer Scholz Edler von Rarancze (Oak Leaves on 12/03/44) (Swords on 08/08/44)
30/01/44	SS-Sturmbannführer Bunse
12/03/44	SS-Sturmbannführer Krügel (Oak Leaves on 16/11/44)
12/03/44	SS-Untersturmführer Langendorf
12/03/44	Obersturmbannführer Lohmann
12/03/33	SS-Hauptsturmführer Saalbach
12/03/44	SS-Sturmbannführer Schulz-Streeck
12/03/44	SS-Obersturmführer Seebach
12/03/44	SS-Obersturmbannführer Stoffers
16/06/44	SS-Hauptsturmführer Hämel
11/07/44	SS-Unterscharführer Christophersen
23/08/44	SS-Sturmbannführer Bachmeier
23/08/44	SS-Oberscharführer Hektor
23/08/44	SS-Obersturmbannführer Kausch
05/09/44	SS-Brigadeführer Ziegler (Oak Leaves on 28/04.45)
23/10/44	SS-Hauptsturmführer Gürz
23/10/44	SS-Unterscharführer Sporck

16/11/44 SS-Obersturmbannführer Knöchlein
16/11/44 SS-Hauptsturmführer Lüngen
26/12/44 SS-Obersturmbannführer Karl
26/12/44 SS-Sturmbannführer Potschka
28/02/45 SS-Obersturmführer Rott
16/03/45 SS-Hauptsturmführer Vogt (Oak Leaves)
20/04/45 SS-Obersturmführer Hund
29/04/45 SS-Obersturmbannführer Gieseler

16th SS Panzer Grenadier Division "Reichsführer-SS"
(16. SS-Panzergrenadier-Division "Reichsführer-SS")

On 7 April 1941 Hitler ordered the formation of an operations staff for the coming war against the Soviet Union. It was to be responsible for combating communist partisans and elements of the population in the rear areas of the conquered territories in the east. In the beginning, the bulk of the attached units were members of reinforced SS Death's Head Regiments (*SS-Totenkopf Standarten*), which had already performed occupation duties.

On 15 May 1941 an escort battalion (*Begleit-Bataillon*) was formed in Oranienburg. This motorized unit was intended to perform the role of reconnaissance battalion. Under the command of *SS-Sturmbannführer* Schützeck, it consisted of:

1st Company
Motorcycle Platoon
Armored Car Platoon
Anti-Tank Platoon (37-mm anti-tank guns)
2nd Company
3 rifle platoons with trucks
3rd Company
2 platoons of 20-mm anti-aircraft guns]
1 platoon of 37-mm anti-aircraft guns

In July 1941 the companies moved to Training Camp "Arys," on Spirding Lake in East Prussia, for further training—live firing with anti-tank and anti-aircraft guns. From there, on 3 September 1941 the battalion followed the 2nd SS Infantry Brigade, which since the end of July had been engaged in combing the forests along the Rositten – Leningrad road. Five days later, after a march of about 900 kilometers, the units reached L Army Corps.

The 1st Company of the Escort Battalion "Reichsführer-SS" was equipped with the 37-mm Pak 35/36 anti-tank gun (top right). This weapon was incapable of knocking out a T-34; however, it was effective against infantry targets using HE ammunition. Here members of the battalion pose with a captured Soviet tank.

The corps had just breached the Stalin Line near Luga, taking unbelievable casualties in the process. After the arrival of the 2nd SS Infantry Brigade, the Battle of Krasnogvardeysk—an important transportation junction outside Leningrad—began. While the SS Police Division captured the city five days later, the Escort Battalion "*Reichsführer-SS*" was employed to eliminate a pocket of enemy troops near Vyritsa (about 30 km southeast of Krasnogvardeysk). During the fighting members of the battalion shot captured Red Army soldiers.

When the situation at the front became more critical, the battalion's attachment to the Operations Staff "*Reichsführer-SS*" ended. At the beginning of October 1941 it occupied positions in the main line of resistance in the Uritsk – Pushkin area within the 2nd SS Infantry Brigade. When the brigade was sent to XXVIII Army Corps, the Escort Battalion "*Reichsführer-SS*" went too. Attached to the 122nd Infantry Division, from 3 to 21 November the unit took part in defensive fighting in the Kolpino area on the Neva River (approx. 12 km east of Pushkin). After *SS-Sturmbannführer* Schützeck was wounded, at the beginning of December *SS-Sturmbannführer* Garthe[17] took over the battalion. There was heavy fighting in the Krasny Bor area until 16 December. Consisting mainly of 18-year-old boys, the battalion was pulled out of action and spent Christmas in Riga. On 28 December 1941 the men marched back to the front—this time to XXXVIII Army Corps (16th Army), which was positioned on the Volkhov River north of Lake Ilmen between Myasny-Bor and Spaskaya-Polist (approx. 30 km north of Novgorod).

There the Escort Battalion "*Reichsführer-SS*" was combined with the Volunteer Legion "*Flandern*" and SS Flak Battalion "*Ost*" to form SS Battle Group Debes, which was employed as XXXVIII Army Corps's tactical reserve.

When, on 13 January 1942, the Red Army succeeded in breaking through the positions of the XXXVIII Army Corps, SS Battle Group Debes was ordered to establish a defensive front along the road from Novgorod to Leningrad. On 28 January 1942 the units were attached to Infantry Regiment 424 (126th Infantry Division). About two weeks later the fighting temporarily died down. *SS-Sturmbannführer* Burk took over the SS units under *SS-Oberführer Debes* which, still attached to the 126th Infantry Division, were deployed near Bol'shoye Zamosh'ye. The severe cold — temperatures reached minus 51 degrees C—was responsible for more casualties than enemy action.

After the enemy assault bogged down in front of Lyuban, on 15 March 1942 the Germans launched a counteroffensive. The German units succeeded in cutting off the Soviet 2nd Shock Army on the Volkhov from its own lines, temporarily at first. There was fierce, back-and-forth fighting, and the Red Army was able to reestablish contact. Not until the end of May 1942 did Army Group North succeed in closing the pocket again—this time for good.

Troops in action in northern Russia during the winter of 1941-42.

In June 1942 the German forces set about destroying the surrounded enemy units. The Escort Battalion "*Reichsführer-SS*" delivered a decisive blow on 15-16 June, which forced the stubborn enemy back across the main road into the swamps around Kerstiy Brook. Four days later the decimated battalion was pulled out of action. It was first sent to Widminnen (east of Lötzen) and finally to Training Camp "Arys." There the men were given their annual leave. The commander of the 2nd SS Motorized Infantry Brigade said goodbye to the men with the following words:

"After three-quarters of a year, the Escort Battalion "Reichsführer-SS" is leaving the brigade, of which it has been a part since leaving Training Camp "Arys."

The battalion proved itself in battle wherever it was deployed; this is true of the anti-partisan operations, initially in the forest and swamps, and soon in the ice and

Actions in northern Russia.

Im Namen des Führers
und Obersten Befehlshabers
der Wehrmacht

verleihe ich

dem

SS-Strm.

Wilhelm B a r e n s c h e e r

1./Begl.Btl.RFSS

das

Eiserne Kreuz 2.Klasse

...Div.Gef.St......, den 2. M ä r z 19 42

Leutnant und Div.-Kdr.

(...nstgrad und Dienststellung)

One received the Iron Cross – the other a wooden cross.

snow of the Russian winter, as well as in the 122nd Infantry Division's trenches on the Neva before Leningrad, and finally in the fighting on the important road from Pogostye to Olonne. But every man of the battalion and also every member of the 2nd SS Infantry Brigade will remember with special pride the battalion's performance in the fighting on the Novgorod to Leningrad road. Arriving at a desperate time, it halted the Russian penetration in a fierce struggle at great sacrifice and never yielded a foot of ground despite bitter cold, despite massive Russian superiority and despite the desperate situation. Only those who took part in these battles can appreciate what the battalion accomplished in those weeks and months.

It fills me with satisfaction that, just before leaving the brigade, the battalion had the privilege of delivering the decisive blow on 15 and 16 June 1942, driving the stubborn-fighting Red enemy trapped in the Volkhov pocket across the Novgorod – Leningrad road into the swamps around Kerstiy Brook, where he now faces ultimate destruction. There, too, the battalion demonstrated its offensive spirit and will to win and showed that, as proper SS men, every member of the battalion is ready to give his best and his last in the struggle for the future of the German people.

It is with sincere regret that the brigade sees this magnificent battalion leave its midst, and from its heart it wishes it and its proven commander all the best and soldier's luck for its new assignments in the future."

At the end of July 1942 the battalion was sent to the SS and Police Commander "Zhitomir," *SS-Oberführer* Hellwig, for anti-partisan operations in the Pripyat Marshes. The units combed the vast areas of forest and swamps southeast of Mozyr several times by the end of November 1942. The battalion was subsequently transferred back to SS Training Camp "Arys" for the formation of the SS Assault Brigade "*Reichsführer-SS*" (*SS-Sturmbrigade "Reichsführer-SS"*). The later *SS-Oberscharführer* Köck recalled:

"*I was born in the Sudetenland in 1921. After the annexation I moved to Nuremberg. There, in the summer of 1939, I had to report for a weekend of military training. I arrived in my Sunday suit. Instructors from the 21st Infantry Regiment chased us all over the fields, and more than my Sunday suit was ruined. I had no desire to do that again, and so when I read a recruiting poster for the 9th Infantry Regiment in Potsdam, I immediately signed up. Unfortunately, at that time the regiment had all the men it needed. Less than enthused about the weekend meat grinder, in the Pirckheimerstraß I read recruiting posters for the Waffen-SS. I volunteered on the spot, and on 1 May 1940 I was called up by the Leibstandarte Adolf Hitler in Berlin-Lichterfelde. Soon afterwards, I and about 150 other recruits formed a training company at the SS officer's school in Brunswick. There we were the guinea pigs of the SS officer cadets. In May 1941 I was transferred to Oranienburg to join the signals platoon of the newly-created escort battalion. Before the Russian campaign we were sent to Allenstein. There we lost our first man. A dispatch rider, who like most of our drivers had just received his driver's license, crashed into a tree. We got rid of the stress of night exercises by bathing in the many Mazurian lakes.*

On 22 June 1941 we set off after the fighting units. After fierce fighting before Leningrad, on 16 December 1941 we were pulled out to Riga. On 28 December we were placed on alert again. We were first sent to Kingisepp, where we found temporary quarters in a wrecked barracks. In January 1942 we became involved in heavy fighting on the Chudovo – Leningrad road in the Koptsy area. The Russians attacked by night and we counterattacked by day. In the severe cold everything froze—the bread, the sausage, and the tea.

Not far from us there were Spanish and Flemish troops. They appeared to fight well, but the language barrier prevented us from communicating with them. I do remember, however, that the Spaniards were very fond of red wine!

In August 1942 we were sent to Zhitomir to fight partisans. The partisans rarely showed themselves, but they often plundered small villages and stole cattle, grain, and eggs. In one village we found the bodies of 25 civilians killed by the partisans.

As there were plenty of fish in the Pripyat, we went fishing with hand grenades and improved our menu. One day our commander, SS-Obersturmbannführer Gesele, gave me the job of giving Hitler's pilot, Flugkapitän Hans Bauer, an hour and a half of infantry training. Probably just in case. Finally we returned to Arys-North."

The Assault Brigade "Reichsführer-SS" (*Die Sturmbrigade "Reichsführer-SS"*)

At Himmler's suggestion, on 14 February 1943 Hitler authorized the expansion of the 823-man-strong Motorized Escort Battalion "*Reichsführer-SS*" into a new-style assault brigade. It was organized as follows:

brigade headquarters
grenadier battalion (motorized)
heavy battalion (motorized)

SS-Obersturmbannführer Gesele was responsible for formation of the new unit at Training Camp "Rennes" in the area of the Commander-in-Chief West. Erich Rommel recalled:

"In December 1942 I was transferred to the Escort Battalion "Reichsführer-SS" at Arys-North. Our company was simply called "Circus Hatz" after our company commander, SS-Obersturmführer Hatz. Company formation in ranks, left turn march, practice alerts at night, or falling out at night with a straw mattress, crawling in and singing a song were the order of the day.

At Arys I had a terrible experience which I cannot forget. While assigned to the guard detail at the company office an SS-Sturmmann had opened a parcel and eaten the contents. He was court-martialed and was shot in front of the entire battalion, which was assembled in a U-shaped formation. That was a shock for us 17 to 18 year old boys. I couldn't talk about it for a long time!

We were all hungry, and sometimes we could buy a loaf of bread from the customs officer. Each morning one had to show his squad leader a slice of his bread. Twice I was unable to do so, and each time SS-Unterscharführer Vogel gave me a slice of his bread. Then I began fetching the food for the NCOs and in the morning the coffee. In return I was allowed to eat whatever was left over.

As a member of a mortar team I had to carry the base plate, plus the bipod or tube. I became the 2nd mortar man, then the 1st. Just before Christmas 1942 we were issued complete winter outfits (kidney protector, felt-lined boots, fur-lined jacket, and snow smock) and trained hard. It was said that we were headed for southern Russia to help in the relief of Stalingrad. Our time was filled with maneuvers with live ammunition, but also parade marches between Arys North and South (18 km).

Then everything changed again: we had to turn in our winter equipment and in minus 30 degree temperatures we were loaded onto trains. Eight days later we arrived in Brittany, between Rennes and St. Brieux. The violets were blooming and it was already quite warm. In addition to more training (including house fighting), we also performed occupation-related tasks: standing guard on the Channel, capturing shot-down enemy airmen, and preventing and combating sabotage."

While ethnic German personnel and equipment to expand the brigade were arriving in France, it was becoming obvious that the Allies were planning a landing in Italy. After a special briefing at the beginning of June 1943 the OKW war diary noted:

"*The enemy is continuing his air attacks against airfields and ports in Sicily, Sardinia, and Italy in undiminished strength with powerful units. This must be seen as preparations for a landing operation.*"

As part of the German countermeasures, on 20 June 1943 the assault brigade, which was still in the formation process, was transported by rail to Massa – Carrara. From there it was moved to Corsica, in part by aircraft. The SS men were to establish a bridgehead in the southern part of the island for the imminent evacuation of German forces from Sardinia. The headquarters took up quarters in Sartene. An SS war correspondent described the situation on Corsica on 25 August 1943:

"*On 19 and 20 August 1943 the Commander-in-Chief South, Generalfeldmarschall Kesselring, paid Corsica a visit, which we were able to capture on film. As yet there hasn't been any fighting, apart from a minor night action when the enemy dropped supply containers with weapons and ammunition. 108 American-made machine-guns and submachine-guns were captured along with 100,000 rounds of ammunition.*

The brigade had moved into its quartering area in southern Corsica at the beginning of August; brigade headquarters was initially quartered in Aullene, but this mountain village had too few buildings and a few days later we moved to Sartene.

The heat remains almost unbearable and is causing many brush and grass fires, so that in the evening it looks as if the mountains are burning… Our rations are still inadequate. One cannot buy potatoes here, at best grapes, green figs, almonds, a few apples. Olive oil is rarely seen and is very expensive; almost the same can be said of crayfish and lobster. In general everything is very expensive, and we lack the means to purchase.

The people here are just as lazy as those in Russia, the houses are dirty, the living quarters have only a few pieces of poor furniture. The people spend most of the day sitting in front of their houses. Half-wild pigs and goats roam the streets. Small donkeys are typically used as riding and pack animals."

When Italy dropped out of the war the numerically-weak brigade was placed in a precarious situation. The withdrawal of the 90th Panzer Grenadier Division from Sardinia was carried out as planned, but only because the 70,000 Italian soldiers there remained neutral. Two days later the division took over the bridgehead position in the south and the Assault Brigade "*Reichsführer-SS*" was ordered by the Wehrmacht Commander of Corsica to take Bastia, which had been fortified by the Italian troops. SS units occupied the town on 13 September after heavy fighting. Thus the evacuation of the German troops was assured. The Bastia bridgehead was abandoned on 3 October 1943. At 21:00 hours the last German soldiers left the island in boats for Livorno. Three days later the Wehrmacht communiqué declared:

"*An SS assault brigade distinguished itself in the fighting on the island of Corsica.*"

The assault brigade's commander, *SS-Obersturmbannführer* Gesele, was awarded the Knight's Cross of the Iron Cross. Josef Köck recalled:

"*In France we conducted an exercise: defending against an airborne landing by the British. We concentrated on the "paratroopers" dropped from the air and not until too late did we realize that they were straw dummies. General Students parachute troops, who had landed in gliders, then attacked us from the rear. They also captured all our vehicles. All in all a most embarrassing performance by our officers...*

The western allies often dropped weapons and equipment for the resistance at night. We discovered an entire ammunition dump in a monastery! We were then transported by rail from Loudeác to Massa and Carrara. From Pisa we flew in Ju 52s to Bastia, on Corsica. At first we camped in tents, left our three 88-mm anti-aircraft guns there and marched farther south. There we covered the crossing by the 90th Panzer Grenadier Division, which had been newly formed on Sardinia. When the Italians broke their alliance with Germany in September 1943 many Italian soldiers threw their weapons away, drank wine, and celebrated without end. We fell back in the direction of Bastia, which had meanwhile been occupied by Italian troops, and were fired on by our own three 88-mm flak! Our comrades had been barricaded in the citadel in Bastia. I was ordered to call in the air force to bomb Bastia, but we were able to occupy the port city, making the raid unnecessary. Atmospheric interference made it impossible to cancel the bombardment, and so the next morning German aircraft bombed the city. I was actually supposed to have been court-martialed over this, but then it wasn't called. We made it back to Livorno by ferry."

In mid-December 1941 Herbert Garthe was promoted to SS-Sturmbannführer and assumed command of the Escort Battalion Reichsführer-SS. Garthe's final command was SS Panzer Grenadier Regiment 35 as an SS-Obersturmbannführer.

SS officers Hummel, Klinzmann and Hoffstädt (from l. to r.).

The 16th SS Panzer Grenadier Division "Reichsführer-SS" (*Die 16. SS-Panzergrenadier-Division "Reichsführer-SS"*)

On 23 September 1943 Hitler ordered the creation of a central reserve which, formed using members of the 1926 age class, was to consist of 14 divisions (10 infantry, 2 parachute, and 2 SS divisions). One of the two SS divisions was to be created in the Laibach area by reorganizing and strengthening the assault brigade. As establishment of the division did not begin until mid-November 1943, the assault brigade temporarily remained in the Livorno area.[18]

The new division was to be organized as a panzer grenadier division:

Division Headquarters
SS Panzer Grenadier Regiment 35 (I – III Battalion)
SS Panzer Grenadier Regiment 36 (I – III Battalion)
SS Artillery Regiment 16 (I – III Battalion)
SS Panzer Battalion 16
SS Reconnaissance Battalion 16
SS Assault Gun Battalion 16
SS Pioneer Battalion 16
SS Flak Battalion 16
SS Medical Battalion 16
SS Signals Battalion 16
SS Economic Battalion 16
SS Field Replacement Battalion 16

SS-Brigadeführer Simon was named division commander. While most of the approximately 1,800 members of the assault brigade stayed near Livorno, the following were initially sent into the formation area at Laibach:

500 recruits from SS Panzer Grenadier Training and Replacement Battalion 1
1,000 recruits from SS Panzer Grenadier Training and Replacement Battalion 2
500 recruits from SS Panzer Grenadier Training and Replacement Battalion 3
500 recruits from SS Panzer Grenadier Training and Replacement Battalion 9
500 recruits from the SS Flak Training and Replacement Regiment
2,000 recruits from the SS Artillery Training and Replacement Regiment
300 recruits from SS Signals Training and Replacement Battalion 1
200 recruits from SS Pioneer Training and Replacement Battalion 2
1,200 recruits from the SS Special Purpose Training Battalion
800 recruits from the SS Driver Training and Replacement Battalion

Max Simon commanded the 16th SS Panzer Grenadier Division from October 1943 to October 1944.

SS-Oberführer Otto Baum took over the division on 24/10/44 and led it until the end of the war.

Cuff title of the 16th SS Panzer Grenadier Division "Reichsführer-SS."

The bulk of the replacements were ethnic Germans from the 1926 age class. About 3,000 were from the 1902 to 1922 age classes. By 31 December 1943, at which time the division was directly attached to the 14th Army, its strength had already risen to:

	Officers	NCOs	Enlisted Men	Total
Actual	203	1,191	11,326	12,720
	1.6%	9.4%	89%	100%
Authorized	551	3,230	11,368	15,149
	3.6%	21.4%	75%	100%

In addition to shortages of materiel, about six weeks after formation began the division was short about 2,300 officers and non-commissioned officers.

When, on 22 January 1944, Anglo-American forces landed near Nettuno, the elements of the former Assault Brigade "*Reichsführer-SS*" stationed in the Livorno area were bolstered by the II Battalion, SS Panzer Grenadier Regiment 36, which had been formed in the Laibach area. The combined force marched into the area southwest of Cisterna with a strength of:

Officers	NCOs	Enlisted Men	Total
22	173	1,530	1,725
1.3%	10%	88.7%	100%

As the enemy landing force failed to exploit the element of surprise, instead taking its time to assemble in the beachhead, the weak German forces were able to establish a defensive front. Attempts were made to smash the beachhead on 16 and 29 February 1944, however, these failed because of the enemy's naval artillery and inadequate German forces, resulting in an initial period of positional warfare.

On 20 March 1944 the battle group, commanded by *SS-Sturmbannführer* Knöchlein (commander SS Panzer Grenadier Regiment 36), was briefly reinforced by a battalion of Italian SS volunteers. Then, on 18 April it handed over the positions along the bridgehead to Grenadier Regiment 1028 (motorized) and returned to the 16th SS Panzer Grenadier Division *"Reichsführer-SS."*

At the beginning of March 1944 the division had marched out of the formation area at Laibach to Baden, near Vienna, and was attached to LVIII Army Corps for the occupation of Hungary. On 19 March 1944 the division, bolstered by the SS Panzer Grenadier Instruction Regiment (*SS-Panzergrenadier-Lehr-Regiment*) and a newly-formed Escort Battalion *"Reichsführer-SS"* (*Begleit-Bataillon "Reichsführer-SS"*), crossed the Hungarian border into the Raab area. After the creation of the so-called "Operations Area East Hungary" on 24 March 1944, the

In June 1943 the unit was transported by rail to Italy…

and in March 1944 to Hungary ("Operation Margarethe").

16 SS Panzer Grenadier Division was ordered into the Debrecen area to serve as an occupation unit. Josef Köck, then an *SS-Oberscharführer*, recalled:

"The Allies landed at Nettuno on my birthday. We were immediately alerted and marched from Banja Luka to Cisterna. Battle Group Knöchlein arrived from Laibach to reinforce us. We were positioned at the bridgehead south of the Pontine Marshes. The German forces there were so weak that the Americans could have advanced without difficulty. The enemy's naval artillery was so powerful that we became terribly frightened. One day, when Tiger tanks advanced into the marches on the only road, the naval gunfire tossed them into the air almost as if they were toys. After just under two months I got away from the beachhead and returned to Burgenland before 19 March 1944. There I visited my wife and small daughter. The march into Hungary was unopposed; however, as leader of the signals platoon I lost two of my three Kfz. 17s. It was very cold at night; the coolant froze and the engine blocks cracked. During our stay in Debrecen we often went swimming in the local thermal bath, which was very pleasant given the winter-like temperatures."

On 31 March 1944 the division was widely dispersed: SS Battle Group Knöchlein was still at the Nettuno bridgehead and approximately 50% was in the Baden assembly area, while the rest was in the Raab area or already in the new quartering area west of Debrecen. At that time the division's total strength was:

	Officers	NCOs	Enlisted Men	Total
Actual	335	1,602	14,912	16,849
	2%	9.5%	88.5%	100%
Authorized	551	3,230	11,368	15,149
	3.6%	21.4%	75%	100%

The bulk of the recruits had at least an approximately three-month period of basic training behind them, and after the delivery of the necessary equipment unit training could be started at a training camp. The division was still far from being ready for operational use, however, as it had only the following heavy weapons:

	Actual	Authorized
Sturmgeschütz III, 75-mm	12	73[19]
Armored troop carriers	--	32
Kettenkräder	--	42
Motorcycles	42	643
Cars (off-road capable)	489	341
Cars	49	778

The commander of SS Pioneer Battalion 16, SS-Sturmbannführer Lange, inspects the assembled battalion in Hungary on 20/4/44. To his left is a Honvéd officer.

Maultiere (halftrack truck)	--	31
Trucks (off-road capable)	52	833
Trucks	442	796
Prime movers (1 – 5 tons)	22	171
Prime movers (8 – 18 tons)	12	50
Raupenschlepper Ost	13	2
Panzerabwehrkanonen 40 (75-mm anti-tank guns)	24	40
Heavy infantry guns	7	12
Artillery pieces	40	40
Machine-guns	994	959

On 1 April 1944 the division commander, *SS-Brigadeführer* Simon, assessed his unit as follows:

"*Training[has been] impossible for four weeks because of the Hungary action, resuming this week.*

The division is spread over Europe, from the Mediterranean to the Carpathians.

30% motorized, the rest immobile.

The division is not operational, cannot be used for any combat assignment."

On 16 April 1944 Himmler ordered the division to transfer 4,500 men with light and heavy infantry weapons to the badly-battered 3rd SS Panzer Division "*Totenkopf*" in Rumania. As replacements, the 16th SS Panzer Grenadier Division "*Reichsführer-SS*" was to be sent the first 2,000 newly-inducted ethnic Germans from Hungary, with the recruit classification *"kv-SS"* (meeting SS standards). The recruiting of ethnic Germans was part of the agreement between Germany and Hungary reached on 14 April 1944. As the division's actual strength had been greater than its authorized strength, this personnel reduction of about 2,500 men did not result in any restrictions. It did have a negative impact, however, in that 4,500 trained men were released and replaced by ethnic-German recruits. Some of these men, who began arriving on 15 May 1944, had not even been issued uniforms. As a result of this move the division's employability was restricted for more months.

On 13 May 1944 Anglo-American forces launched an offensive east of the Gulf of Gaeta into the area northwest of Cassino, and the Wehrmacht Operations Staff feared a major landing on the Ligurian coast. On 18 May 1944 the 16th SS Panzer Grenadier Division received marching orders for the Livorno area in Italy. The division entrained on 20 May and was declared OKW reserve, although it could only be used in the event of an enemy landing. The first elements of the division—SS Armored Reconnaissance Battalion 16 and the division headquarters—arrived in Livorno in the area of LXXV Army Corps on 26 May. On 31 May the main body of the division occupied the sector between Marina di Carrara and Calafuria.

In order to reinforce the units for the coming mission, the new Escort Battalion "*Reichsführer-SS*"—already attached—and the SS Panzer Grenadier Instruction Regiment were integrated into the division in exchange for the exhausted SS Battle Group Knöchlein.[20] At the same time SS Assault Gun Battalion 16 was renamed SS Anti-Tank Battalion 16 and reorganized:

1st Company (12 75-mm anti-tank guns, towed)
2nd Company (10 assault guns)
3rd Company (12 20-mm anti-aircraft guns on self-propelled carriages)

Just five weeks later, the Allies has pushed the German troops back about 270 km to the north and reached the line Grosseto – Perugia. The 16th Panzer Grenadier Division was placed on alert on 18 June 1944.

On 29 June the enemy reached the Cecina area. The day before XIV Panzer Corps had assumed command there, including over the 16th Panzer Grenadier Division, which had been sent into the main line of resistance. The division occupied positions near Cecina along the river of the same name. Josef Köck recalled:

"*We returned to Italy in May 1944. In the Villa Reale, north of Lucca, radio operator Braune accidentally shot a comrade in the belly while cleaning his weapon. He was sentenced to 20 years imprisonment and was supposed to be sent to Danzig-Matzkau. He remained with us at first, however, and was given permanent mine-seeking duty. In the course of his duties he was buried alive, and afterwards suffered a nervous breakdown. I don't know what became of him. In this area our anti-aircraft guns also shot down an Me 109, having assumed it to be an English machine. My men and I were sent into the vineyards to look for the supposed Tommy. When we found him he shouted at us, calling us asses. On 30 June 1944 I was in an accident while driving in the serpentines. We had to drive without lights, and so I left the road and overturned. I ended up in the division aid station, which was in a church, with a fractured skull and broken ribs. There were already 200 wounded comrades there. Finally, on 17 August 1944 I left the hospital for the SS Signals Replacement Battalion in Nuremberg. This ended my association with the 16th SS Panzer Grenadier Division "Reichsführer-SS.""*

After the Commander-in-Chief Southwest succeeded in halting the enemy along the Cecina and went over to an organized defense, it was possible to begin work on the so-called "Green Position" between La Spezia and Pesaro. This in-depth position in the mountains was supposed to permanently stop any further advance.

The enemy knew how difficult it was to attack a mountain position in autumn or winter and tried to reach the Green Position before summer ended by committing vast quantities of materiel. On 6 July 1944, in extraordinarily fierce fighting, the

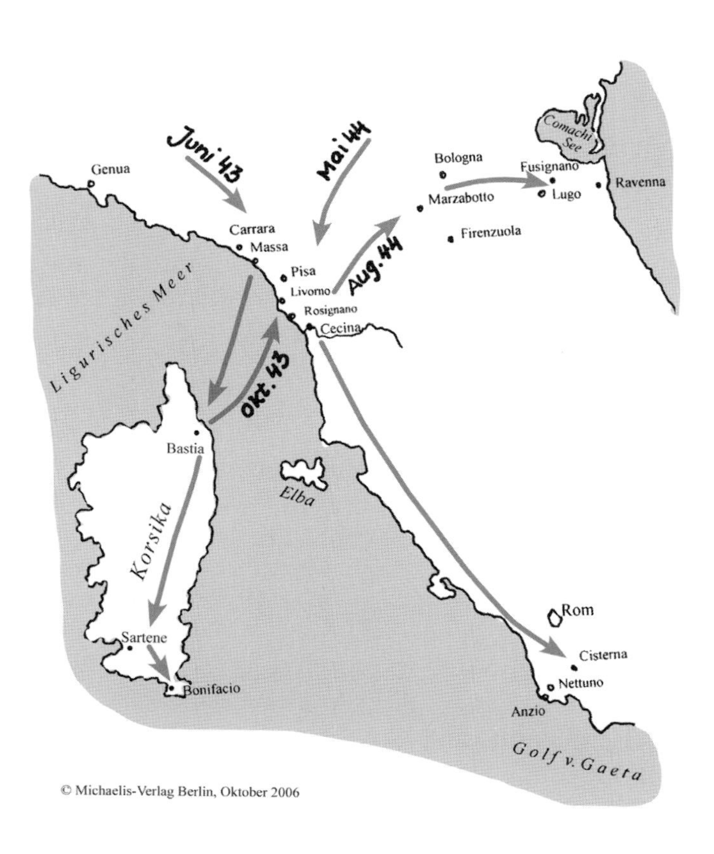

zugleich Personalausweis

Nr. 135

für

den SS Pz Gren.

(Dienstgrad)

ab 1-9-43 (Datum) SS Sturmmann (neuer Dienstgrad)

1.11.43 SS-Rottenführer

Georg Nicolescu

(Vor- und Zuname)

Beschriftung und Nummer der Erkennungsmarke II/Ndl SS Pz Gren Rgt 2 Nr. 181

Blutgruppe AB

Gasmaskengröße 2

Wehrnummer St. aus Rum.

K 167 SS-Vordruckverlag W. F. Mayr, Miesbach 17384

geb. am 26-12-1919 in Bukarest (Ort, Kreis, Verw.-Bezirk)

Religion r.k. Stand, Beruf Student

Personalbeschreibung:

Größe 168 Gestalt schlank

Gesicht oval Haar dunkelblond

Bart keinen Augen blau

Besondere Kennzeichen (z. B. Brillenträger): —

Schuhzeuglänge 43 Schuhzeugweite 4 1/2

Georg Nicolescu

(Vor- u. Zuname, eigenhändige Unterschrift des Inhabers)

Die Richtigkeit der nicht umrandeten Angaben auf Seite 1 und 2 und der eigenhändigen Unterschrift des Inhabers bescheinigt

den 12. AUG. 1943

5./Ndl. SS-Pz. Gren. Rgt. 2

(Ausfertigende Truppenteil, Dienststelle)

SS-Ostuf. u. Kp.-Führer

(Eigenh. Unterschrift, Dienstgrad u. Dienststellg. d. Vorges.)

Dienststelle Feldpostnummer

2

über die Richtigkeit der Zusätze und Berichtigungen auf Seite 1 und 2

Lfd. Nr.	Art der Änderung	auf Seite	Datum	Truppenteil Unterschrift	Dienstgrad und Dienststelle
1	Ernennung	1	1.9.43		SS-Ostuf. u. Kp.-Führer
2	Dienstgrad	1	1.11.43		SS-Obersturmf. u. Kp.-Führer

3

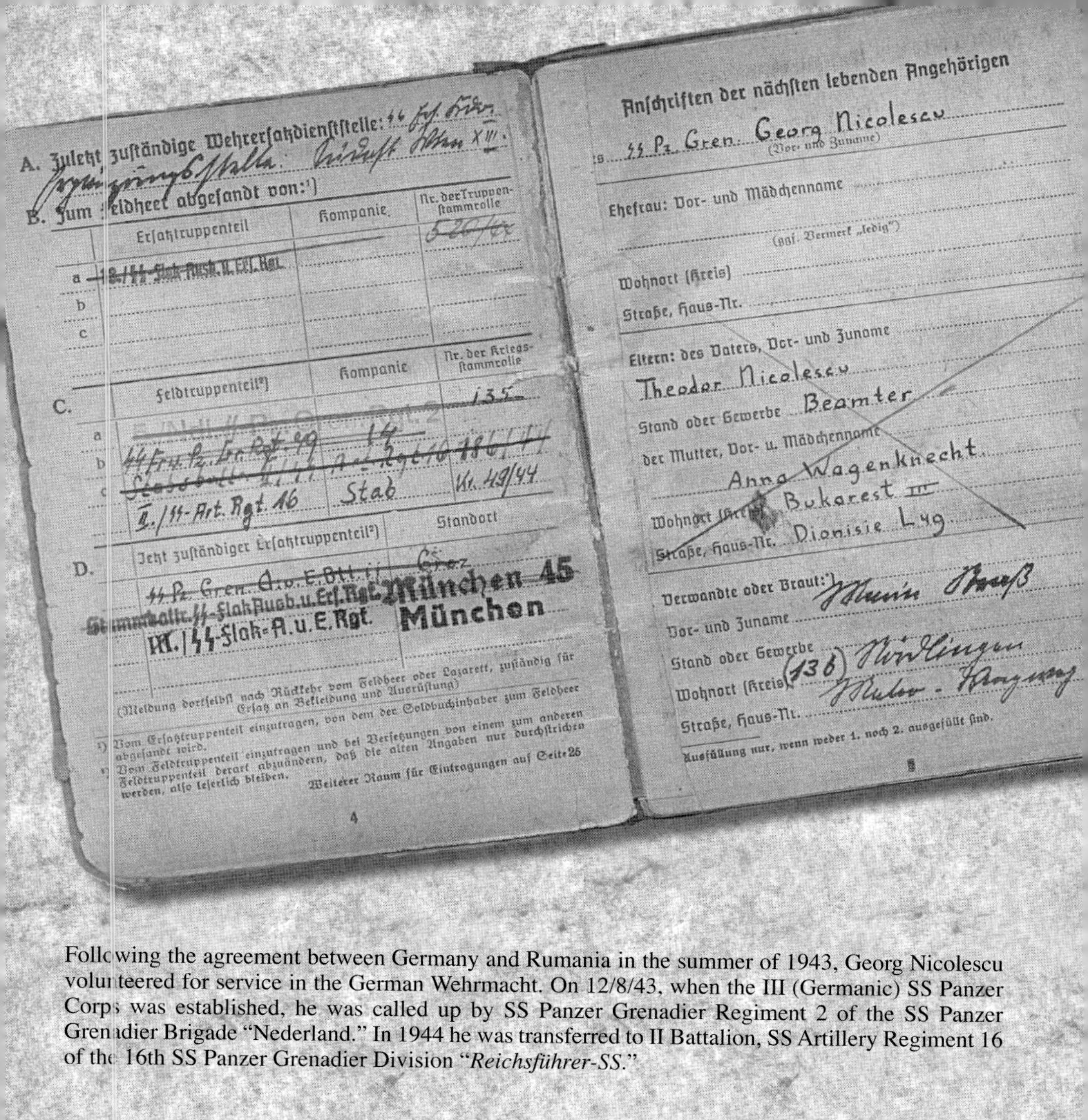
A. Zuletzt zuständige Wehrersatzdienststelle: SS Erg. Amt Ergänzungsstelle Südost Wien XVII.

B. Zum Feldheer abgesandt von:[1]

	Ersatztruppenteil	Kompanie	Nr. der Truppen-stammrolle
a	~~18./SS-Flak-Ausb. u. Ers. Rgt.~~		~~520/44~~
b			
c			

C.

	Feldtruppenteil[2]	Kompanie	Nr. der Kriegs-stammrolle
a			135
b	~~SS Frw. Pz. Gr. Rgt. 49~~	~~14~~	~~186/44~~
c	~~Stabsbttr. II./SS-Art. Rgt. 16~~		
	II./SS-Art. Rgt. 16	Stab	Kt. 49/44

D.

Jetzt zuständiger Ersatztruppenteil[2]	Standort
~~SS Pz. Gren. A. u. E. Btl. 11~~	~~Graz~~
~~Stammabtlg. SS-Flak-Ausb. u. Ers. Rgt.~~	~~München 45~~
III./SS-Flak-A. u. E. Rgt.	München

(Meldung dortselbst nach Rückkehr vom Feldheer oder Lazarett, zuständig für Ersatz an Bekleidung und Ausrüstung)

[1] Vom Ersatztruppenteil einzutragen, von dem der Soldbuchinhaber zum Feldheer abgesandt wird.
[2] Vom Feldtruppenteil einzutragen und bei Versetzungen von einem zum anderen Feldtruppenteil derart abzuändern, daß die alten Angaben nur durchstrichen werden, also leserlich bleiben.

Weiterer Raum für Eintragungen auf Seite 25

4

Anschriften der nächsten lebenden Angehörigen

des SS Pz. Gren. Georg Nicolescu (Vor- und Zuname)

Ehefrau: Vor- und Mädchenname

(ggf. Vermerk „ledig")

Wohnort (Kreis)

Straße, Haus-Nr.

Eltern: des Vaters, Vor- und Zuname Theodor Nicolescu

Stand oder Gewerbe Beamter

der Mutter, Vor- u. Mädchenname Anna Wagenknecht

Wohnort (Kreis) Bukarest III

Straße, Haus-Nr. Dionisie L. 49

Verwandte oder Braut:[1]

Vor- und Zuname

Stand oder Gewerbe

Wohnort (Kreis) (13b) Nördlingen

Straße, Haus-Nr.

[1] Ausfüllung nur, wenn weder 1. noch 2. ausgefüllt sind.

5

Following the agreement between Germany and Rumania in the summer of 1943, Georg Nicolescu volunteered for service in the German Wehrmacht. On 12/8/43, when the III (Germanic) SS Panzer Corps was established, he was called up by SS Panzer Grenadier Regiment 2 of the SS Panzer Grenadier Brigade "Nederland." In 1944 he was transferred to II Battalion, SS Artillery Regiment 16 of the 16th SS Panzer Grenadier Division "*Reichsführer-SS*."

enemy broke into the positions of the 16th SS Panzer Grenadier Division in the Rossignano area. In about 14 days the division lost:

3 battalion commanders
26 company commanders
18 platoon leaders

Total casualties during the period from 30 June until the fighting retreat beyond the Arno in the Pisa area on 19 July 1944 were:

Officers	NCOs	Enlisted Men	Total
51	328	1,607	1,986

On 20 July 1944 the Wehrmacht communiqué declared:

"*Under the command of SS-Gruppenführer und Generalleutnant der Waffen-SS Simon, the 16th SS Panzer Grenadier Division Reichsführer-SS has distinguished itself through its steadfastness and bravery during the heavy fighting on the Ligurian Coast.*"

The heavy casualties of almost 2,000 men were due in part to the stubbornness with which the units held their positions, but also to the inadequate training and lack of combat experience of much of the division.

Troops of the U.S. 5th Army reached the Arno on 23 July 1944. Attempts to cross to the north bank of the river failed and the front was stabilized. On 3 August 1944 the 16th SS Panzer Grenadier Division was pulled out of the defense and moved to the area south of Bologna. There the unit was to rest and receive the equipment it had been missing since the turn of the year 1943-44.

Orders were issued for a general evacuation of the civilian population to protect the lines of communication behind the front. Every civilian subsequently found in the area was viewed as a suspected partisan or partisan sympathizer. Since September 1943 attacks by partisans against the German military had become almost daily events. German troops were sniped at and ambushed, and—as in the Balkans—the violence escalated.

In August 1944 the 16th SS Panzer Grenadier Division carried out an anti-partisan operation. According to the 14th Army's daily report of 26 August 1944 a total of 1,480 partisans, partisan supporters, and suspected partisans were arrested and 323 men were "*killed in battle*." On 12 August 1944 elements of SS Armored Reconnaissance Battalion 16 under the command of *SS-Sturmbannführer* Reder shot about 130 civilians in the north Italian town of Sant'Anna di Stazzema.

After 17 members of the division were killed in a partisan ambush, on 20 August 1944 members of SS Armored Reconnaissance Battalion 16 burned down 16 houses in Bardine San Terenzo and shot a total of 35 residents. Günther Wick, then an *SS-Rottenführer*, recalled:

"I was a member of SS Armored Battalion 16, and in August 1944 we were stationed between Pisa and Lucca to guard the river crossings. Because of transmission trouble I had to go to the workshop company, which was below Fosdinovo, about 20 km behind the front. The driver and the radio operator had to stay with the vehicle to help with repairs. We figured that we would have it easy for about a week. Next door to the workshop company were the headquarters company and the train, consisting mainly of the slightly wounded and sick. While things were quite relaxed at the front, in the rear order again reigned. The headquarters company had to fall in every evening to receive orders, and the senior NCO assigned the guard and work details for the next day. He concluded by saying: I need 17 men for an escort detachment tomorrow. And to our great astonishment the entire company volunteered. I subsequently engaged the comrades in conversation and asked why they had all volunteered for the mission. We learned the following:

Every fifth day two trucks drove to a central depot in Bologna to fetch everything the armored battalion needed, apart from fuel and ammunition. The drive began while it was still dark to avoid the attentions of the American fighter-bombers, for the enemy had total air superiority and fired on anything moving on the roads. When the trucks reached the depot they were loaded and parked. The soldiers were then free to do as they wished. They could go to the soldier's hostel for something better to eat, they could go to the barber, and afterwards watch a movie in the military theater. The return trip only began after dark. For the men, the interlude in Bologna was like a day's leave. Now we knew why everyone had volunteered.

The next day began quietly, as usual. At about 10 o'clock there was noise in the headquarters company. The men were ordered to fall out: two trucks pulled up, the men got in, and they drove away. What was happening? The word quickly spread. The two trucks on their way to Bologna had been ambushed! They returned at about 14:00 hours. Everyone ran to meet them, everyone wanted to know what had happened. Not a word; everyone stayed silent as the bodies of the 17 comrades were unloaded. They had all been horribly mutilated. I was only able to immediately identify two of the bodies lying there. One was the transport commander. He could be identified by his officer's uniform, and the other comrade was wearing a gauze bandage around his neck, as he had had a carbuncle removed from his neck three days earlier.

None of the 17 comrades had a face any longer! What had happened? A rock slide had forced the trucks to stop at a sharp bend in the road. As the others began

After recovering from a serious wound, in March 1944 Günther Wick was transferred to SS Panzer Battalion 16 in Hungary. He had previously served in SS Panzer Regiment 3 "Totenkopf."

Italy, summer 1944.

clearing away the rocks one of the drivers took advantage of the stop and ran behind a bush to relieve himself. He was about to return when all 17 men were cut down by submachine-gun fire. Then the partisans emerged from cover, picked up heavy stones, and smashed the face of each soldier! That was on 17 August 1944 near Bardine!"

At the beginning of September 1944 the division's strength was:

	Officers	NCOs	Enlisted Men	Total
Actual	368	2,089	12,226	14,683
	2.5%	14.2%	83.3%	100%
Authorized	557	3,388	12,593	16,538
	3.4%	20.5%	76.1%	100%

It relieved the 362nd Infantry Division in the main line of resistance north of Pistoia in the sector of XIV Panzer Corps (14th Army). At the end of the month elements of the 16th SS Panzer Grenadier Division, plus Flak Regiment 105 and IV (East) Battalion, Grenadier Regiment 1059, took part in an anti-partisan operation in the rear area between the Reno and Setto Rivers southeast of Marzabotto. The force lost a total of seven killed and reported 718 enemy dead. This casualty ratio suggests that it could not have been regular combat.

On 24 October 1944 *SS-Oberführer* Baum assumed command of the division. The 16th SS Panzer Grenadier Division's front remained relatively quiet until the beginning of December 1944. Attempts by the enemy to advance on Bologna were repulsed. On 20 December 1944 the units were pulled out of the main line of resistance and, as the Commander-in-Chief Southwest's "*most battle-worthy unit,*" it was initially moved to the area south of Lake Comacchino on the 10th Army's left wing in the role of reserve. On 28 December the division's reported strength was:

	Officers	NCOs	Enlisted Men	Total
Actual	395	2,262	11,566	14,223
	2.8%	15.9%	81.3%	100%
Authorized	557	3,388	12,593	16,538
	3.4%	20.5%	76.1%	100%

That day the division relieved the 98th Infantry Division in LXXIII Army Corps. The positions ran behind the Senio between Lugo and Fusignano west of Ravenna.

In January 1945 Hitler planned an offensive to secure the oil region in Hungary, which was vital to the war effort. The 6th SS Panzer Army was pulled out of the west, and units from Italy, including the 16th SS Panzer Grenadier Division "*Reichsführer-SS*," were ordered to Hungary. Relieved by the 362nd Infantry Division, at the beginning of February 1945 the unit moved into the Nagykanizsa area under the camouflage designation "*Rest and Reequipment Group 13th SS Division*."[21] As part of the effort to fool enemy intelligence, some of the men had to turn in their pass books and were given passes identifying them as members of the *13. Waffen-Gebirgs-Division der SS "Handschar."* They also had to remove the collar patches bearing the *Sig-Runen* and some wore a fez.

After having been constantly on the defensive in Italy, the units were hastily trained in offensive tactics for the coming attack. For about two weeks the division conducted day and night exercises in battalion, regiment, and division strength—some with live ammunition.

Initially employed as OKW reserve, on 1 March 1945 the "*Rest and Reequipment Group 13th SS Division*" was attached to the 2nd Panzer Army in the Nagybajom area. The German command assessed the 16th SS Panzer Grenadier Division as "*conditionally suitable for offensive tasks.*"

"Operation Spring Awakening" began on 6 March 1945. Now attached to the LXVIII Army Corps, the "*Rest and Reequipment Group 13th SS Division*" attacked

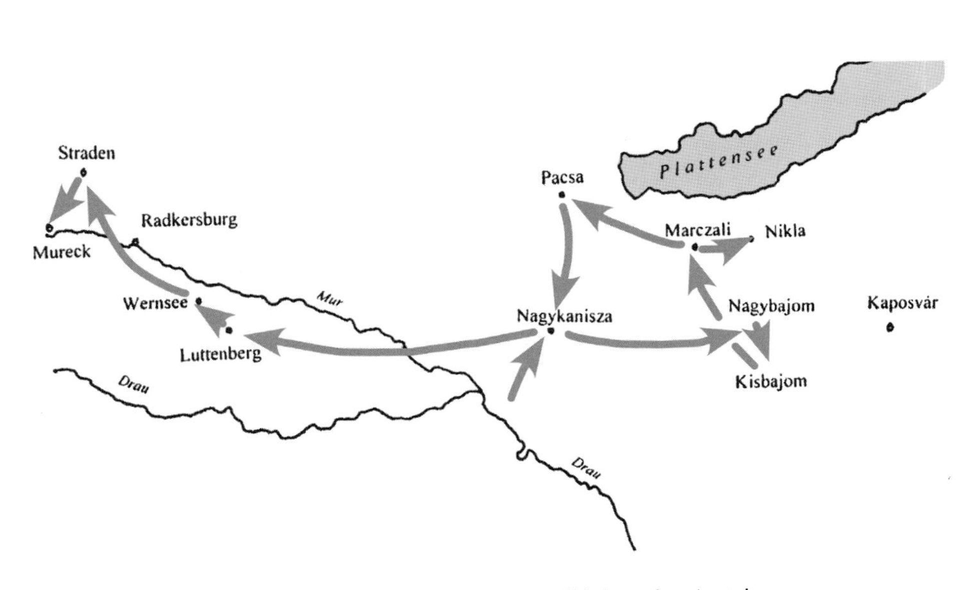

Operations in Hungary and the Withdrawal to Austria

In Italy the men lived in their tents for long periods of time.

The later 16th SS Panzer Grenadier Division "*Reichsführer-SS*" had been equipped with tropical uniforms since its deployment to Corsica.

from the center of Nagybajom toward Kaposvár; however, the attack bogged down in the face of the enemy's in-depth system of positions. The units gained just a few hundred meters of ground before they were halted by Soviet counterattacks.

Having failed to achieve any progress, on 10 March 1945 Army Group South shifted its main effort farther south toward Kisbajom. On that day the "*Rest and Reequipment Group 13th SS Division*" submitted a strength report:[22]

1 strong, 3 average, and 3 weak (grenadier) battalions
1 average field replacement battalion
1 strong pioneer battalion
20 heavy anti-tank guns
19 assault guns
6 light and 3 heavy batteries of artillery
2 heavy flak batteries

There were only Bulgarian troops around Kisbajom and they offered little resistance; consequently, the attack gained about 4 kilometers of ground and reached Kisbajom. There too, however, the attack bogged down. The corps reorganized its forces and orders were given for the "*Rest and Reequipment Group 13th SS Division*" to march into the Marczali area. Attached to XXII Mountain Corps, from there it was to advance along Lake Balaton toward I Cavalry Corps.

On 14 March 1945 the division launched an attack toward Nikla which was unsuccessful. The unit, which on 18 March 1945 was again designated 16th SS Panzer Grenadier Division "*Reichsführer-SS*," suffered heavy casualties in its attacks on the enemy's in-depth defense system. On 24 March it reported a rations strength of 9,399 and a combat strength of 3,134 men. Since 6 March the division had lost about 4,000 men killed, wounded, or captured.

Three days later the unit was ordered to assemble in the Pacsa area (about 15 km west of Lake Balaton) as the 2nd Panzer Army's mobile reserve. The Lake Balaton offensive had been called off. On 29-30 March 1945 the Red Army launched a major offensive against the 2nd Panzer Army, and the 16th Panzer Grenadier Division was split. Attached to various units, the main body withdrew toward the Nagy-Kanizsa area. Fighting its way back through the Mur bridgehead at Alsólendva toward the "Reich Defense Position," the division's movements sometimes resembled a hectic flight. A former member of the division recalled:

"The entire division moved to Hungary in the second half of January 1945 and an action began on 6 March 1945. When the attack bogged down our division commander, SS-Oberführer Baum, came forward in a Schwimmwagen, upbraided

our commanding officer (SS Panzer Battalion 16), and ordered the attack to continue at full steam. The town was taken—among those killed were SS-Oberscharführer Keun and SS-Hauptsturmführer Paul. In a subsequent night action my squad lost SS-Sturmmänner Rolf Hoffmann and Ludwig Möller and SS-Grenadier Walter Abt. Ultimately there were only 8 men left from the original 55 of my platoon. The others had been wounded or killed. Twice we had received replacements from other SS units—mostly recruits."

On 4 April 1945 the division was assembled in the Luttenberg area at the disposal of the XXII Mountain Corps and was ordered to destroy the Soviet bridgehead near Wernsee. Three days later it was moved into the threatened Straden area and attached to the I Cavalry Corps.

At Training Camp "Alt-Neudorf" the division's strength was bolstered by the addition of members of the air force and navy plus Hungarian volunteers. On 13 April 1945 the division occupied assembly areas in the Radkersburg area, and in the days that followed it was deployed north and west of Straden to north of Poppendorf.

The premature surrender by Army Group C in Italy resulted in 2nd Panzer Army's rear communications being placed in jeopardy. To resolve this situation, the 16th SS Panzer Grenadier Division was relieved in its positions and attached to XXII Mountain Corps. Loaded onto trains in Leibnitz – Mureck, elements saw action in the Marburg – Lavamünd area. Elements of the division surrendered in the Klagenfurt area and the area west of Graz. The then *SS-Rottenführer* Bickelmann remembered:

"Born on 27 March 1921, I began my service in the Waffen-SS on 30 January 1941 at the age of 19. I received my initial training with the SS Signals Replacement Battalion Unna, and on 1 May 1941 I was transferred to the SS Flak Battalion "Reich." From 2 June to 10 December 1944 I served in the cadre company of the SS Signals Replacement Battalion 1in Eichstätt. Then, on 11 December 1944 I was transferred to SS Panzer Grenadier Regiment 36. There I was always up front with the infantry with my backpack radio. On the day that Hitler committed suicide we were in Styria. The name of the village was Gnas. The military police dragged young men out of barns and houses and hung them as traitors. The English were already very close. Our battalion stayed in Austria, and a few days later we were taken prisoner in Molzsbichl, near Spittal on the Drau. Everything descended into chaos in the final days of April 1945. The first to leave were our officers. No one wanted to be in charge any more. From the prisoner of war camp in Tarento (Italy), I returned home on 28 March 1947."

Although it achieved full strength in personnel, Germany's materiel situation prevented the division from becoming fully equipped with weapons and equipment. Not until about 18 months after its establishment could the division be reported "conditionally suited for offensive missions." Nevertheless, the 16th SS Panzer Grenadier Division "*Reichsführer-SS*" was considered a relatively potent formation in a defensive role in Italy, a secondary theater. The division achieved no success in the Lake Balaton offensive due to the Red Army's massive superiority in defense and offense.

Military Postal Numbers

Division Headquarters	36 800
SS Panzer Grenadier Regiment 35	38 204
I Battalion	37 347
II Battalion	01 011
III Battalion	36 124
SS Panzer Grenadier Regiment 36	40 006
I Battalion	42 040
II Battalion	41 196
III Battalion	38 030
SS Artillery Regiment 16	40 590
I Battalion	43 115
II Battalion	43 905
III Battalion	00 818
IV Battalion	42 357
SS Panzer Battalion 16	39 202
SS Repair Battalion 16	46 462
SS Armored Reconnaissance Battalion	39 610
SS Assault Gun Battalion 16	37 196
SS Pioneer Battalion 16	45 312
SS Flak Battalion 16	44 214
SS Medical Battalion 16	41 456
SS Signals Battalion 16	41 688
SS Economic Battalion 16	43 495
SS Division Supply Services 16	45 992
SS Field Replacement Battalion 16	44 848

Commanding Officers

10/43 – 10/44 SS-Gruppenführer Simon
10/44 – 04/45 SS-Oberführer Baum

Wearers of the Knight's Cross of the Iron Cross

04/07 SS-Obersturmbannführer Gesele

17th SS Panzer Grenadier Division "Götz von Berlichingen"

17. SS-Panzergrenadier-Division "Götz von Berlichingen"

After orders were issued for the formation of the 9th and 10th SS Panzer Grenadier Divisions in autumn 1942 for the creation of a central reserve, on 23 September 1943 Hitler ordered the formation of a second central reserve. Created using members of the 1926 age class, it was to consist of 14 divisions (10 infantry, 2 parachute, and 2 SS divisions).

The two SS divisions were the 16th and 17th SS Panzer Grenadier Divisions. Hitler ordered the establishment of the latter on 30 October 1943.

The division units were to be established in the Saumur/Loire – Bressuire – Parthenay – Poitiers – Chatellerault area in the 1st Army's sector:

Division Headquarters
SS Panzer Grenadier Regiment 37 (I – III Battalion)
SS Panzer Grenadier Regiment 38 (I – III Battalion)
SS Artillery Regiment 17 (I – III Battalion)
SS Armored Reconnaissance Battalion 17
SS Assault Gun Battalion 17[23]
SS Flak Battalion 17
SS Pioneer Battalion 17
SS Signals Battalion 17
SS Medical Battalion 17
SS Economic Battalion 17
SS Supply Services 17

Swearing-in.

Basic (individual) training was completed by 15 February 1944. According to plan squad training was completed by 15 March, and platoon and company training by 15 April 1944. Battalion, regiment, and division training was to follow under the supervision of the division commander, *SS-Oberführer* Ostendorff.

In addition to 270 instructors from the SS non-commissioned officer's school in Posen-Treskau, the following 17-year-olds joined the division:

200 recruits from SS Panzer Grenadier Training and Replacement Battalion 2
1,000 recruits from SS Panzer Grenadier Training and Replacement Battalion 4
1,000 recruits from SS Panzer Grenadier Training and Replacement Battalion 12
300 recruits from SS Infantry Gun Training and Replacement Battalion 1
300 recruits from SS Signals Training and Replacement Battalion 5
200 recruits from SS Pioneer Training and Replacement Battalion 1
300 recruits from the SS Special Purpose Training Battalion "Debica"

The remaining personnel were mostly ethnic Germans from Rumania (age classes 1902 to 1922):

500 recruits from SS Panzer Grenadier Training and Replacement Battalion 1
500 recruits from SS Panzer Grenadier Training and Replacement Battalion 3
500 recruits from SS Panzer Grenadier Training and Replacement Battalion 5
500 recruits from SS Panzer Grenadier Training and Replacement Battalion 9
200 recruits from SS Signals Training and Replacement Battalion 1
300 recruits from SS Pioneer Training and Replacement Battalion 1

Personnel were also transferred from the 10th SS Panzer Grenadier Division "*Frundsberg*" which, together with the 17th SS Panzer Grenadier Division "*Götz von Berlichingen*," would later form the VII SS Panzer Corps.

After the assignment of additional ethnic German personnel, by 31 December 1943 the unit was already able to submit the following strength report:

	Officers	NCOs	Enlisted Men	Total
Actual	142	817	10,188	11,147
	1.3%	7.3%	91.4%	100%
Authorized	358	3,412	12,636	16,406
	2.2%	20.8%	77%	100%

In January 1944 the elements of the division assembled at the French training camp "Thouars" (located between Saumur and Parthenay) for the completion of basic training and the switch to squad training.

The division command was to carry out the training of about 500 drivers itself, using requisitioned French vehicles.

On 10 April 1944 *Reichsführer-SS* Himmler and *SS-Oberstgruppenführer* Dietrich visited the division, which had just completed platoon and company training. Himmler gave a stirring speech in which he gave the division the name "Götz von Berlichingen,"[24] which had been bestowed upon it by Hitler:

"*My dear guests,*

Herr General von Geyr,

Herr Oberleutnant von Berlichingen,

my old friend Sepp Dietrich and you, my men of the 17th newly-formed SS Panzer Grenadier Division, who today have been awarded the name Götz von Berlichingen!

Your commander, SS-Brigadeführer Ostendorff, has just personally assured me that you all are prepared to do your duty always, true to the great example you assume with this name.

I am happy that I can be with you on this day, and I am pleased to be able to say a few words to you on the day this cuff title is bestowed upon you [...] The title "Götz von Berlichingen" has been awarded to you by the Führer, and today you receive the cuff title bearing his name. He lived from 1480 to 1562. It is sometimes shameful how little the German people know of their own history and their great men. Of Götz von Berlichingen, every innocent boy and every foolish man will say to you: "Yes, the famous quotation of Götz von Berlichingen." Beyond that, people know as little about him as they do about Frundsberg or Prince Eugen. You, who belong to the division upon which the Führer has bestowed this name, you who should wear this name with pride, you must know more about him. You must know why he was great.

I would like to begin with his loyalty, which he upheld in every case in which he promised it. He was a prime example of the loyal retainer, who stood up for his sovereign and was ready to give his life for him. [...] The second thing I would like to emphasize is his mailed fist—not because of the iron in his fist, but because of his unbridled fighting spirit, which found its expression in his refusal to yield. From an inner calling he was a warrior and fighter and soldier. To be a soldier, that is a calling. He who has it cannot escape it, he fights on even if he has just one arm, one eye, or one leg. […] I now come to the third and perhaps the most important thing in this time of war: we are in the fifth year of this conflict, of this Second World War, this struggle against plutocracy and bolshevism, with Slavs and West Europeans and Americans. This struggle is nearing its climax. The time is coming when one of these three cannot go on […] Recently the Japanese proudly announced that in two years of warfare they have not lost even 400 men as prisoners. A proud confession, a proud declaration. We cannot match their numbers. Today we are around 300,000 German SS men, with at least the same number of volunteers, Norwegians, Danes, Bosnians, Galicians, Estonians, Latvians, Italians, Walloons, and French, a large number of men we have recruited to fight for Germany, in whom we have instilled loyalty to the Führer Adolf Hitler and to whom we have taught order, soldierly values, and fighting skills. We German SS men can say of ourselves: I believe, after critical examination, that we have lost scarcely more than 300-400 men as prisoners.

I expect one thing of this Götz von Berlichingen Division, that the thought of giving in, either in the spiritual or political realm, will never enter the mind of even one man. But above all I expect one other thing, that every member of this division, down to the newest recruit, will be taught to accept, and believe that even for the individual there can never be a surrender, never can one allow oneself to be captured. For my sake, never misuse the famous quotation in this division. You must and should use it in just one case: when the foe or the greatest enemy a man has, his inner coward, forces you to decide whether you should give in, whether you should yield, without the express order to retreat, then from commander to the youngest man, to the 17-year-old recruits, you should use the phrase: "You can kiss our ass!"

By acting thus, you will hold firm against yourself and others, then you will stand, fight, and win. Fight for this, live for this, win for this: Heil Hitler! And now I will award to your commanders, who will in turn award to you, the cuff bands bearing the title **Götz von Berlichingen***. Wear them with pride, live and fight true to the example set by the division's soldierly forefather."*

At the end of April 1944 the "almost complete" unit was made available to the OKW as a reserve force for the expected invasion, and when the Allies landed in Normandy on 6 June the division's almost 18,000 men were transferred from the OKW reserve in the quartering area south of the Loire to the 7th Army. The next day the division, less SS Anti-Tank Battalion 17 and SS Flak Battalion 17, which were still in the formation process, set out for II Parachute Corps' sector, about 210 kilometers away.

The commanding general of the I SS Panzer Corps, *SS-Obergruppenführer* Sepp Dietrich, receives *Reichsführer-SS* Himmler. Below: Dietrich in conversation with SS-Oberführer Ostendorff.

Reichsführer-SS Himmler salutes the commander of Panzer Group West, General der Panzertruppen Geyr von Schweppenburg. On the left is SS-Oberführer Ostendorff, commander of the 17th SS Panzer Grenadier Division "Götz von Berlichingen," 10 April 1944.

SS-Oberführer Ostendorff in conversation with General Geyr von Schweppenburg. On the left is the Senior SS and Police Commander "France," *SS-Gruppenführer* Oberg. Center: SS-Oberführer Baum.

The commander of the 2nd SS Panzer Division "Das Reich," SS-Oberführer Lammerding, congratulates SS-Oberführer Ostendorff on being named commander of the 17th SS Panzer Grenadier Division.

Inspecting the units.

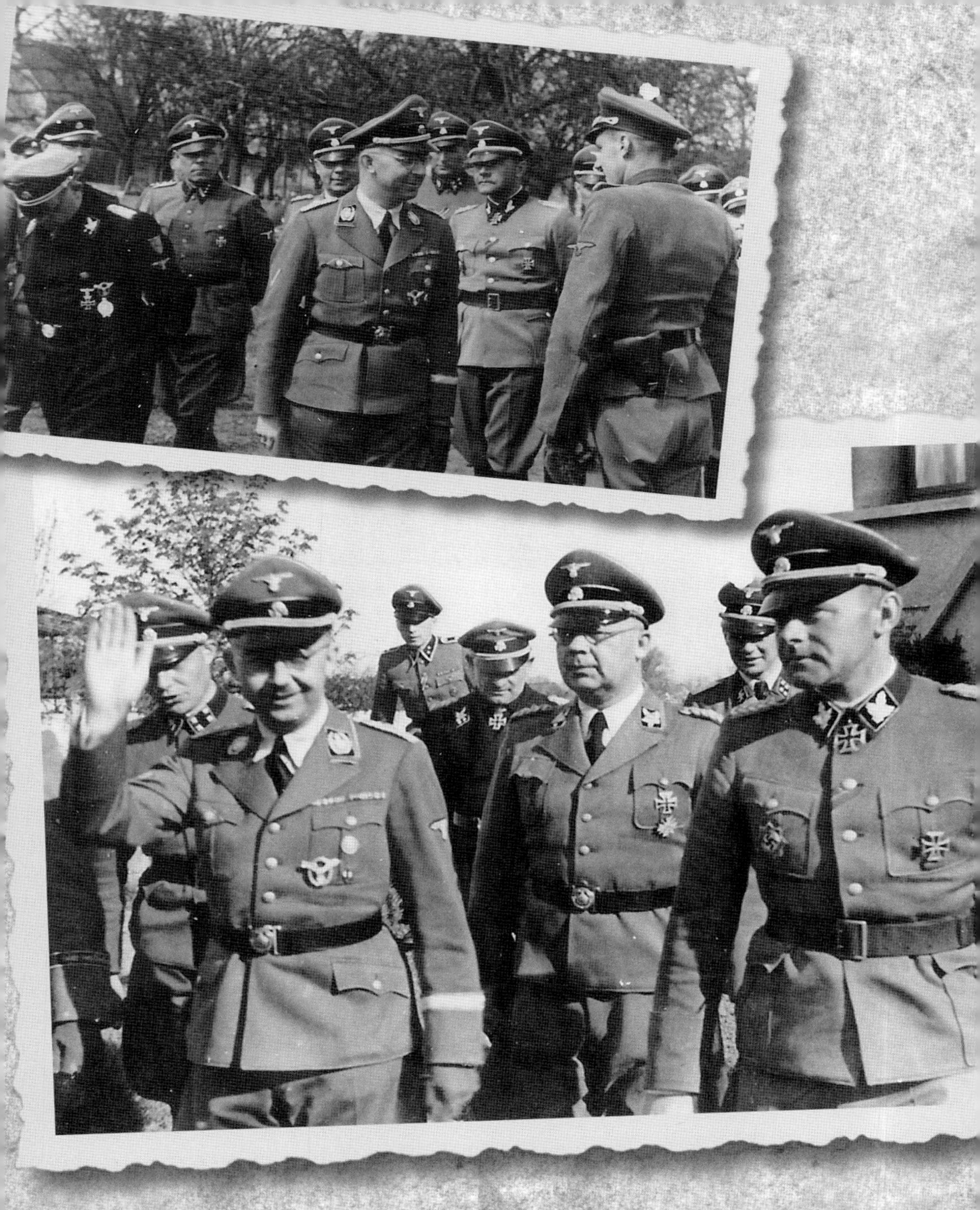

Werner Ostendorff was promoted to *SS-Brigadeführer* on 10/4/44. In the top photo he is still wearing the collar patches of an SS-Oberführer, while in the bottom photo he has the insignia of a *SS-Brigadeführer*.

After passing through Laval, Fougéres, Avranches, Villedieu, and Torigny, the first units arrived in the Balleroy area, which represented the boundary between the American and British invasion sectors. There the division's forces immediately became involved against troops of the American V Corps. When American forces landed near Carentan on 9-10 June, the green 17th SS Panzer Grenadier Division was ordered into the area, which was about 30 kilometers away. The division commander, *SS-Brigadeführer* Ostendorff, was severely wounded in the fierce fighting from 13 to 16 June. On 28 June command of the unit was transferred to *SS-Standartenführer* Binge.

Under constant fire from the enemy's naval artillery and fighter-bombers, until the beginning of July 1944 the units were forced to withdraw toward the area northeast of Perriers and west of St. Lô. On 30 June 1944 the division reported a strength of:

	Officers	NCOs	Enlisted Men	Total
Actual	351	2,029	14,596	16,976
	2.07%	11.93%	86%	100%
Authorized	358	3,412	12,636	16,406
	2.2%	20.8%	77%	100%

After about three weeks in combat the unit was still about 600 men above its authorized strength. It was, however, about 1,400 NCOs under strength, which severely restricted its operational usefulness. As well, with units still being formed, not all of the division's personnel were at the front.

During the course of July 1944 the division's positions were withdrawn to a line approximately 5 kilometers south of the Lessay – Perriers – St. Lô road. This main line of resistance was held until 28 July 1944, when it became apparent that American forces were about to break through the 17th SS Panzer Grenadier Division's lines in the direction of Granville. The troops initially pulled back in a southwesterly direction and then tried to escape to the southeast through gaps in the front. The clogged roads were under constant attack by enemy fighter-bombers, and the division lost much of its heavy weapons and equipment in the process. The division, which had scarcely had an opportunity for unit-level training, had held up under some of the fiercest fighting in the defensive struggle, and on 29 July 1944 it was praised in the Wehrmacht communiqué:

"Under the command of its severely-wounded commanding officer SS-Brigadeführer Ostendorff and his replacement, SS-Standartenführer Baum, the 17th SS Panzer Grenadier Division "Götz von Berlichingen" distinguished itself in

With more than 3,000 17-year-old recruits. 25% of the unit's enlisted men were under 18.

defense and counterattacks during the heavy fighting in the St. Lô – Lessay area in recent weeks."

In addition to the remnants of the 17th SS Panzer Grenadier Division, *SS-Standartenführer* Baum also commanded the deployed elements of the 2nd SS Panzer Division.

On 6 August 1944 the units, which by then were designated only as battle groups, reached the area northeast of Mortain. There, despite the losses sustained in the difficult fighting withdrawal, they were supposed to take part in a counterattack towards Avranches, 34 kilometers away, by XXXXVII Panzer Corps.

A motorized battle group under the command of *SS-Obersturmbannführer* Fick, commander of SS Panzer Grenadier Regiment 37, was formed for the operation. In addition to companies from the regiment, the unit included elements of SS Anti-Tank Battalion 17 and SS Flak Battalion 17, which had arrived. A former member recalled:

"*I am of the 1921 age class, and in 1936 I joined the Hitler Youth cavalry. In 1939 I volunteered for the air force, however, there was a six-month delay because of a move. I didn't want to wait that long to become a soldier, and I thought to*

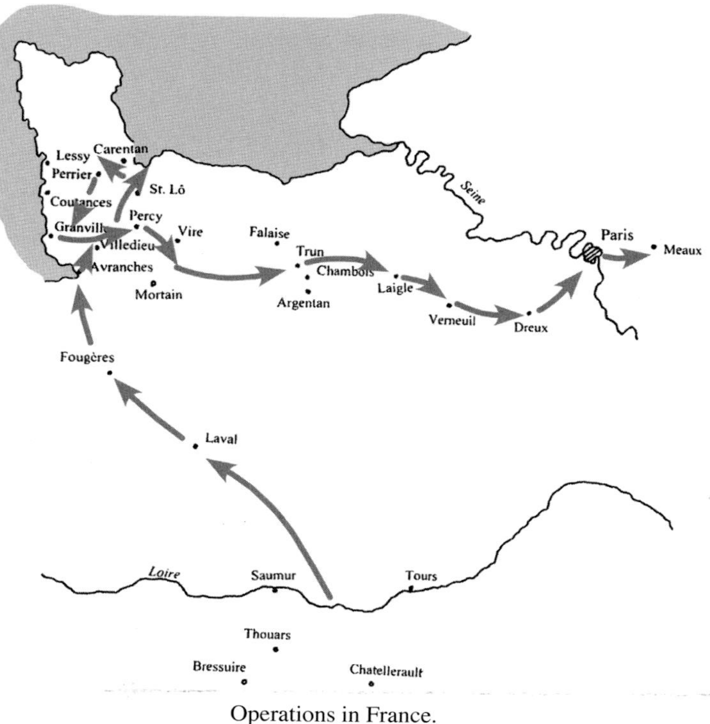

Operations in France.

Camouflage uniforms for the invasion front.

myself, join the SS, that will be quicker. It worked, and in March 1940 I was called up by the SS Artillery Replacement Battalion in Berlin-Lichterfelde. There I earned my driver's license for prime movers, among other things, and I volunteered for the sound-ranging unit. In February 1941 I became a member of the SS Flak Battalion "Das Reich" and remained one until I was wounded near Kharkov in spring 1943. I was resting in a house when suddenly enemy aircraft attacked. A bomb exploded near the house and I was hit in the face by glass splinters from a burst window pane. One went into my left eye and had to be removed in hospital.

After I recovered I went to the SS Flak Replacement Regiment in Arolsen, which soon afterwards was transferred to Munich-Riem. In the meantime I had been ordered to the 3rd course for the wounded at the SS officer's school in Bad Tölz, and from there to the Weapons School in Rerik. After the invasion began we were sent direct to the units without completing our course. For me (SS-Oberjunker) it was the newly-established SS Flak Battalion 17. I travelled by train as far as Paris and then in a double-decker bus by night to Le Mans. There I ran into members of SS Flak Battalion 2, who took me with them to St. Lô. SS Flak Battalion 17 wasn't there yet, as it hadn't been ready to move out. And so I headed south to Saumur on the Loire. I was assigned to the 1st Battery and shot down my first enemy aircraft.

As we had almost no vehicles, we had to use oxen to pull the guns across to the east bank of the Loire. The Amis were advancing steadily and, together with a Wehrmacht battalion, we moved, now with requisitioned vehicles and two Raupenschlepper-Ost, through Tours in the direction of Dijon. The entire time we had no communications with the division!"

"Operation Lüttich" began on the night of 7 August 1944. By that time the 17th SS Panzer Grenadier Division already had two months of major fighting behind it; nevertheless, it was able to encircle American troops on Hill 317 near Mortain. Because of the heavy losses inflicted by enemy fighter-bombers the next day *Generalfeldmarschall* Kluge, the Commander-in-Chief West, called off the operation and ordered the troops to fall back to their starting positions.

On 19 August 1944 the Anglo-American forces succeeded in creating the Falaise Pocket near Chambois. Approximately 80,000 German soldiers were caught, including about 9,000 members of the 17th SS Panzer Grenadier Division who were divided into four battle groups. The Germans launched a breakout attempt on the night of 20 August 1944. The result was apocalyptic conditions: wedged and shot-up columns, and constant fighter-bomber attacks and artillery fire. As a result, only about 20,000 soldiers reached the German lines. Among them were about 1,500 members of the 17th SS Panzer Grenadier Division. The rest were killed or captured.

Initially the remnants of the division marched toward Laigle, approximately 60 km away. From there they reached the Seine by way of Verneuil – Dreux and crossed the river near Paris. There two SS battle groups made up of ethnic Germans were attached to the battered division. For reasons of secrecy they were given the titles 26th and 27th SS Panzer Division.

The 17th SS Panzer Grenadier Division itself was ordered into the Saarbrücken area, where it was to assemble and then rest and reequip. On 8 September 1944 the approximately 4,000 members of the nominal 26th and 27th SS Panzer Divisions and about 2,000 former members of the air force, who had combined to form a battle march unit (*Kampfmarschverband*) at the SS Training Camp "Kurmark," were integrated into the division. After American forces reached Lorraine (Lothringen) in September 1944, various battle groups were dispatched from the assembly area near Saarbrücken to the front, some 60 kilometers away. In the middle of the month the division headquarters had to take over a sector of its own on the Mosel south of Metz within XIII SS Army Corps. While the unit was technically close to authorized strength, its fighting strength was minimal due to its lack of equipment and training. Consequently, on 17 September 1944 the enemy was able to establish a bridgehead in the sector held by the 17th SS Panzer Grenadier Division "*Götz von Berlichingen.*" Without support from the missing heavy weapons (little artillery, no tanks or anti-aircraft guns) counterattacks were unsuccessful.

Together with the elements of the division that had not taken part in the costly actions at the end of July 1944, having instead been assembled in the Le Mans area, the 17th SS Panzer Grenadier Division "*Götz von Berlichingen*" had achieved a strength of 14,000 men. On 29 September 1944 the division submitted the following strength report:

	Officers	NCOs	Enlisted Men	Total
Actual	348	2,820	11,648	14,816
	2.07%	11.93%	86%	100%
Authorized	358	3,412	12,636	16,406
	2.2%	20.8%	77%	100%

Although the division was at 90% strength in personnel, it lacked almost all of the equipment (heavy weapons and vehicles) a panzer grenadier division was supposed to have. A former member of SS Panzer Grenadier Regiment 38 remembered:

"*The composition of our regiment at that time was quite mixed, in keeping with the military situation that then existed. The commanding officer was SS-*

Standartenführer Schützeck. I Battalion was commanded by a police Hauptmann. The company commanders were army Leutnants (Metz officer's school). Most of the NCOs were from the Waffen-SS, a few from the air force. As for the enlisted men, the situation was reversed. The air force men turned SS panzer grenadiers continued wearing their Luftwaffe uniforms with their previous rank badges. Their titles, however, were the same as those of the Waffen-SS. Some men came from the police. Their new ranks were as yet undetermined. They also wore their old police uniforms. Some of the Waffen-SS members wore their camouflage jackets. It was quite a colorful picture.

In October 1944 the regiment was about 20 km behind the main line of resistance for reorganization and training. Although we were panzer grenadiers, I can only remember foot marches. Every week there were three night exercises. Target, direction, and firing practice were carried out by day under good cover. We were very careful not to show ourselves. On 2 November 1944 we marched into our positions. The heavy machine-gun squad and a rifle squad occupied a strongpoint far forward of the actual main line of resistance. It was protected by wooden box mines, captured materiel from Russia. Several bearable misty days were followed by awful rainy days. Our communications trenches and some of our strongpoints were under water. The protective walls in front of our bunkers were inundated. Much disappeared in the water. We were never attacked in our strongpoint, but the Americans broke through on the left and right of Götz von Berlichingen. We abandoned the strongpoints without a fight."

After the front calmed down in October 1944, on 7 November the division commander received an order from the SS-FHA to pull the 17th SS Panzer Grenadier Division out of the front so that it could finally be reequipped. The very next day, however, the Americans resumed the offensive and shattered the division elements positioned around the bridgehead. In the course of the fighting SS Panzer Grenadier Regiment 38, positioned in the north, was separated from the division. A few days later—without having had the opportunity to restore contact with the rest of the division—it was attached to the commander of Fortress Metz. SS Panzer Grenadier Regiment 37 withdrew toward Saargemünd with the other remnants of the division.

The withdrawal by the 17th SS Panzer Grenadier Division enabled the Americans to swing northwards and attack Metz from the south. Meanwhile SS Panzer Grenadier Regiment 38 had been deployed there. After American units reached the outskirts of the city on 18 November 1944, contrary to orders from the fortress commander, regimental commander *SS-Standartenführer* Schützeck ordered his units to leave for the division's new assembly area near Neunkirchen (approx. 16 km NE of Saarbrücken). Schützeck was supposed to have been court-martialled, however, he was killed in action before proceedings could begin. A former *SS-Untersturmführer* recalled the fighting at Metz:

"We passed through Dijon into the area west of Kaiserslautern and finally regained contact with our division. I was placed in charge of an entrenching party. In it were about 100 men from all of SS Flak Battalion 17. Our orders were to dig trenches around Metz. When the Battle of Metz began we withdrew. My men and I were then ordered to provide infantry support for a battery of 88-mm anti-aircraft guns. The next morning, however, the guns were gone. We finally returned to the division area between Saarbrücken and Kaiserslautern. My state of health at that time was very poor due to complete exhaustion and inadequate rations. In February 1945 I was admitted to hospital and later I was assigned to guard duty at the SS officer's school in Tölz."

On 20 December 1944, as per the order issued on 7 November, the division was pulled out of the main line of resistance so that it could be brought up to strength in the Neunkirchen area. What was left of SS Panzer Grenadier Regiment 38, coming from Metz, was absorbed into SS Panzer Grenadier Regiment 37. The SS Panzer Grenadier Instruction Regiment coming from SS Training Camp "Böhmen" formed a new SS Panzer Grenadier Regiment 38. As well as personnel, the division was also issued weapons and equipment. At the end of December 1944 it had the following complement of heavy weapons:

	Authorized	Actual	Enroute
Assault guns	45	17	20
75-mm heavy anti-tank guns	34	7	
Light field howitzers	37	34	
100-mm cannon	4	4	

The division was thus nearing its authorized complement of heavy weapons. At the end of the year it was ordered into the area between Bitsch and Saargemünd. As part of XIII SS Army Corps, the division was supposed to take part in "Operation Nordwind," scheduled to begin on 1 January 1945. On 3 January the Commander-in-Chief of the 1st Army, *General der Infanterie* Obstfelder, called off the offensive. The American's superiority and command of the air placed the attack's objectives far out of reach. A former member recalled:

"Our company, which had just been brought up to strength, attacked Hill 382, general direction of the assault: Groß-Rederching. At dawn the attack bogged down. We had no support from heavy weapons. We lay for hours on the slope with no cover. Very heavy casualties from flanking sniper fire. Finally, after about

The commander of SS Pioneer Battalion 17, SS-Sturmbannführer Fleischer, and members of the battalion.

four hours, three tanks gave us some relief. The Amis withdrew and we hesitantly followed. After about a kilometer a farm appeared before us. Our tanks, advancing from the right, ran into a minefield right in front of the farm. But the attack continued, to the ravine at the railway embankment. There we again suffered heavy casualties to artillery and mortar fire."

On 6 January 1945 American troops broke into the 17th SS Panzer Grenadier Division's positions. In a letter to the head of the SS Operations Headquarters, *SS-Obergruppenführer* Jüttner, Himmler wrote that the SS Panzer Grenadier Instruction Regiment, made up of ethnic German volunteers, had "*run away*" on encountering the enemy despite six months of training.

The front eventually settled down by mid-February. The Americans had tried to break through the 17th SS Panzer Grenadier Division's positions in the direction of Zweibrücken with powerful artillery and air support, however, the units had been able to repulse the attacks in fierce and costly fighting in the Rimlingen area.

On 28 February 1945 the division commander sent the following equipment strength report to the Inspector General of Armored Forces in Berlin:

	Officers	NCOs	Enlisted Men	Total
Actual	393	2,646	12,945	15,984
	2.5%	16.6%	80.9%	100%
Authorized	541	3,170	12,179	16,890
	3.4%	20%	76.6%	100%

	Actual	Serviceable	Under repair
Assault guns	64	41	23
Command tanks	3	1	2
Panzerjäger IV	2	1	1
Bergepanzer III	3	2	1
Panzer 38 (t)	6	5	1
Flakpanzer IV	4	4	
Self-propelled flak	21	19	2

	Actual	Serviceable	Under repair
Light armored cars	6	5	2
Heavy armored cars	6	6	
Light armored troop carriers	2	2	

Heavy armored troop carriers	16	12	4
Light infantry guns	32		
Heavy infantry guns	14		
20-mm anti-aircraft guns	59		
20-mm anti-aircraft guns (quad)	14		
37-mm anti-aircraft guns	18		
88-mm anti-aircraft guns	16		
105-mm light field howitzers	35		
150-mm heavy field howitzers	11		
100-mm field guns	14		
Type 15 machine-guns	318		

The 17th SS Panzer Grenadier Division was relatively well-equipped, apart from the missing armored troop carriers, when the Americans launched their "Undertone" offensive. By 18 March 1945 the German lines had already been forced 15 kilometers toward the north in the direction of Contwig. Though well equipped, it soon became obvious that the division was incapable of standing up to the enormous American pressure.

There followed an attempt, almost a flight, to reach the Rhine near Germersheim by way of Rodalben, Pirmasens, and Landau. When the division crossed the Rhine on 25 March, not only had it lost most of its heavy equipment because of fuel shortages and enemy action, but almost half of its personnel as well. Within a few days efforts to reorganize the shattered division began in the Wiesloch area (approx. 30 km east of the Rhine). After the loss of its heavy weapons the remnants of the division were reorganized as follows:

Division Headquarters
SS Panzer Grenadier Regiment 37 (2 battalions)
SS Panzer Grenadier Regiment 38 (1 battalion)
SS Armored Reconnaissance Battalion 17
Anti-aircraft Artillery Regiment 17 (1 battalion of artillery,
2 battalions in the infantry role)
SS Flak Battalion 17 (infantry role)
SS Panzer Battalion 17 (infantry role)
SS Anti-Tank Battalion 17
SS Pioneer Battalion 17
SS Signals Battalion 17
SS Medical Battalion 17
SS Economic Battalion 17
SS Supply Services 17

SS-Soldbuch
zugleich Personalausweis
Nr. 6465 502
für
den SS-Schütze
(Dienstgrad)
ab ... Datum ... (neuer Dienstgrad)
ab
ab
Otto Kalmbach
(Vor- und Zuname)
Beschriftung und Nummer der
Erkennungsmarke Kr.Ag. SS K.E. A 6465
Blutgruppe A
Gasmaskengröße III
Wehrnummer n.d.
SS-F.H.A. 8. – 50000 – 8. 42.

geb. am 26.4.23 in [illegible] (Ort, Kreis, Verw.-Bezirk)
Bessarabien
Religion ev. Stand, Beruf Bauer

Personalbeschreibung:

Größe 164 Gestalt schlank
Gesicht [illegible] Haar blond
Bart ohne Augen blau
Besondere Kennzeichen (z.B. Brillenträger):
Schuhzeuglänge 42 Schuhzeugweite
Otto Kalmbach
(Vor- u. Zuname, eigenhändige Unterschrift des Inhabers)
Die Richtigkeit der nicht umrandeten Angaben auf Seite 1 und 2 und der eigenhändigen Unterschrift des Inhabers bescheinigt.
den 8. Sep. 1943
SS-Kraftf.-Ausb.- u. Ers.-Abtl.
Stammkompanie
(Ausfertigender Truppenteil, Dienststelle)
(Dienststempel)
SS-Hauptsturmführer u. Kompaniechef
(Eigenh. Unterschrift, Dienstgrad u. Dienststellg. d Vorges.
2

Bescheinigungen über die Richtigkeit der Zusätze und Berichtigungen auf Seiten 1 und 2

Lfd. Nr.	Art der Änderung	auf Seite	Datum	Truppenteil	Unterschrift	Dienstgrad und Dienststellung

3

Otto Kalmbach, an ethnic German from Bessarabia, began his service in the Waffen-SS on 8/9/43.

A. Zuletzt zuständige Wehrersatzdienststelle:

ss Ers. Kdo. Südost, Wien

B. Zum Feldheer abgesandt von:[1])

	Ersatztruppenteil	Kompanie	Nr. der Truppenstammrolle
a	~~ss-Kraftf.-Ausb.- u. Ers.-Abt.~~ Stammkompanie		~~6465~~
b	~~3./ss-Kraftf.-Ausb.- u. Ers.-Abt.~~		~~3/38~~
c	~~ss-Kraftf.-Ausb.- u. Ers.-Rgt. Genesenden-Kompanie~~		~~2571~~
	~~ss-K. A. u. E. A. 4. Komp.~~		~~1018~~

C.	Feldtruppenteil[2])	Kompanie	Nr. der Kriegsstammrolle
a	ss-Div. Sich. Kp. 17		541/45
b			
c			

D.	Jetzt zuständig. Ersatztruppenteil[2])	Standort
	~~ss-Kraftf.-Ausb.- u. Ers.-Abt.~~	~~Weimar-Buchenwald~~
	ss-Pz. Gren. A. u. E. Btl. 17	Iglau

(Meldung dortselbst nach Rückkehr vom Feldheer oder Lazarett, zuständig für Ersatz an Bekleidung und Ausrüstung)

[1]) Vom Ersatztruppenteil einzutragen, von dem der Soldbuchinhaber zum Feldheer abgesandt wird.

[2]) Vom Feldtruppenteil einzutragen und bei Versetzungen von einem zum anderen Feldtruppenteil derart abzuändern, daß die alten Angaben nur durchstrichen werden, also leserlich bleiben.

Weiterer Raum für Eintragungen auf Seite 25.

4

des Otto Kalmbach (Vor- und Zuname)

1. Ehefrau: Vor- und Mädchenname

(ggf. Vermerk „ledig")

Wohnort (Kreis)

Straße, Haus-Nr.

2. Eltern: des Vaters, Vor- und Zuname Jakob Kalmbach

Stand oder Gewerbe

der Mutter, Vor- und Mädchenname

Wohnort (Kreis) Gatzken, Kr. ... Westpr.

Straße, Haus-Nr.

5a

3. Verwandte oder Braut:*)

Vor- und Zuname

Stand oder Gewerbe

Wohnort (Kreis)

Straße, Haus-Nr.

*) Ausfüllung nur, wenn weder 1. noch 2. ausgefüllt sind.

5

Nachw...

Lazarett ... der Laz.-Aufnahme

Reserve-Lazarett ...enberg/Thür. 11. Jan. 1944

21. 8. 44.

31 b

San.-Haupt ...webel

9. 2. 45 – 31 b –

20

... des absendenden Truppenteils (Hauptmann usw. Oberfeldwebel) | Tag und Monat / Jahr der Entlassung aus dem Lazarett | Etwaige Bemerkung in Bezug auf die Entlassung aus dem Lazarett (übergeführt nach NN., als geheilt zum Truppenteil usw.) | Unterschrift des die Entlassung bewirkenden Lazarettbeamten

Hauptfeldwebel 15. 1. 44 Rechnungsführer

21. 9. 44 K.v. E. Truppe

Zahlmeister

16. 2. 45

Mitgegebene Wertsachen und Papiere siehe folgende Seiten!

21

Transferred to SS Division Security Company 17, Kalmbach was wounded by shell fragments in August 1944 and on 9 February 1945.

At the end of March 1945 the reorganized Division was moved into the Neckarelz – Möckmühl area, 30 kilometers away. There, behind the Neckar, the division assumed the main burden of the defense within XIII Army Corps (Wehrmacht). At the beginning of April 1945 the unit, which was being employed purely as infantry, had to be withdrawn farther to the east and occupied temporary positions between Kocher and Jagst.

On 7 April 1945 Army Group G recorded the following strength for the 17th SS Panzer Grenadier Division:

Rations strength:	8,811 soldiers
Fighting strength:	6,389 soldiers
Sturmgeschütze III	12, of which 7 were serviceable
Sturmgeschütze IV	2
75-mm anti-tank guns (S.P.)	18
88-mm anti-aircraft guns (towed)	2

The division thus had only about 1,500 men assigned to the supply units. The number of weapons concentrated in SS Anti-Tank Battalion 17 show that it was the only unit capable of operations. While the unit slowly withdrew towards the east, on 13 April 1945 the OKW ordered the division to move into the Nuremberg area. Given the division's limited mobility (only about 20% of the entire division), the march to LXXXII Army Corps brought serious problems with it. The commanding

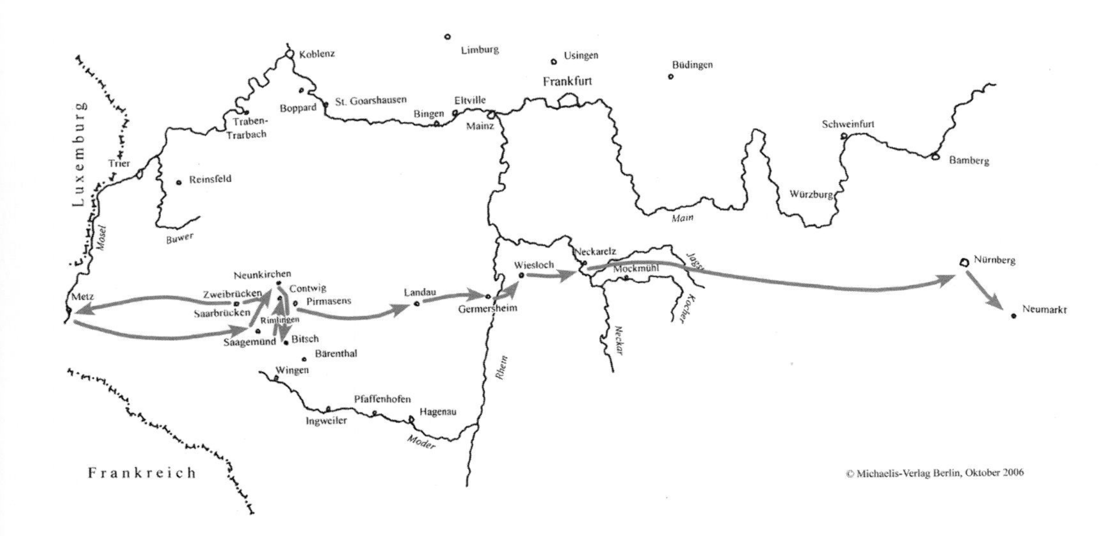

general of XIII Army Corps, *Generalleutnant* Bork, bade the division farewell with the following words:

"In the period from 29/3 to 14/4/45, the 17th SS Panzer Grenadier Division has fought steadfastly and bravely at the Neckar, Jagst, and Kocher. The division bore the main burden of the fighting during that time and played a significant part in the corps' defensive successes. I wish the division good luck in its future missions, which will surely not be easy."

While SS Panzer Grenadier Regiment 38—restored to three-battalion strength, mainly through the addition of air force personnel—was drawn into the fighting at Nuremberg and almost destroyed, the rest of the division arrived in the Neumarkt area on 16 April. As the last remaining artillery battalion had been left behind with XIII Army Corps because of transport difficulties, the two artillery battalions brought in from Beneschau for the 38th SS Grenadier Division were attached to the division.

On 18 April 1945 pressure from the enemy forced the unit to withdraw to the south. Three days later the units were in the Berching area, and on 25-26 April 1945 they reached the Danube near Neustadt. There the division command reported a strength of:

	Officers	NCOs	Enlisted Men	Total
Actual	233	1,418	5,059	6,710
	3.5%	21.1%	75.4%	100%
Authorized	541	3,170	12,179	16,890
	3.4%	20%	76.6%	100%

Ordered into the Dürnbucher Forest as army reserve, the division commander was simultaneously placed in command of the 38th Grenadier Division. By combining all of the men being used as infantry and the troops that had broken through from Nuremberg, SS Panzer Grenadier Regiment 38 was reformed there.

On 27 April 1945 the unit received orders to join XIII Army Corps in the Augsburg area. However, the rapid advance by American armored forces forced the units—which became involved in frequent fire fights—to withdraw in the direction of Ostmark (Austria). There was heavy fighting in the Dachau area on 28 April 1945 and near Penzberg the next day.

In 12 days the division had moved 180 kilometers to the south, constantly in contact with the enemy and subjected to unceasing air attacks. At the beginning of May 1945 the remnants of the division were in the Bad Tölz – Tegernsee – Lenggries area, and from there they marched in the direction of the Achensee. When the surrender came SS Panzer Grenadier Regiment was still north of Kreuth,

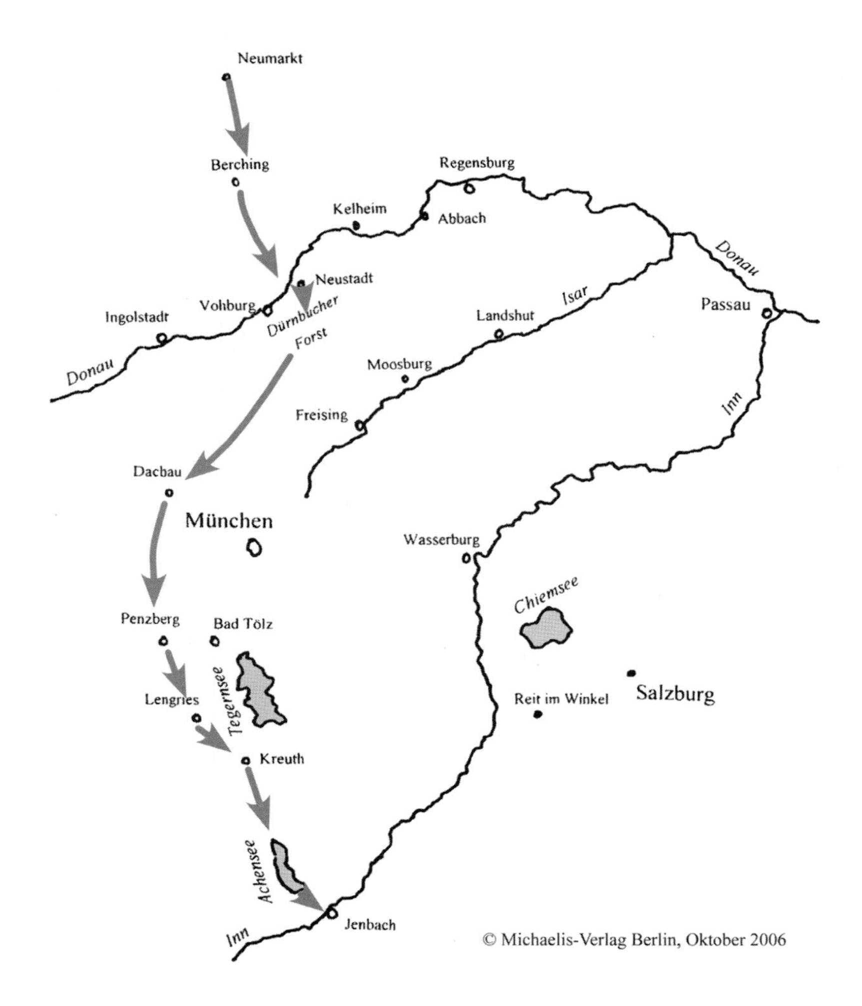

while SS Panzer Grenadier Regiment 38 was already near Jenbach.

This brought to an end the nine-month combat history of the 17th SS Panzer Grenadier Division "*Götz von Berlichingen*." The inexperienced division was badly battered during its first actions, mainly on account of the materiel superiority of the Allies. Until the end of the war there was never an opportunity to completely restore the division's strength or train its replacements, which varied greatly in quality. Heavy weapons and equipment were delivered to the division, but each time they were lost in action.

Military Postal Numbers

Division Headquarters	36 777
SS Panzer Grenadier Regiment 37	32 208
I Battalion	37 044
II Battalion	35 272
III Battalion	44 181
SS Panzer Grenadier Regiment 38	44 261
I Battalion	33 237
II Battalion	45 020
III Battalion	32 974
SS Artillery Regiment 17	39 022
I Battalion	45 144
II Battalion	39 312
III Battalion	30 615
SS Panzer Battalion 17	34 356
SS Repair Battalion 17	41 722
SS Armored Reconnaissance Battalion 17	38 594
SS Assault Gun Battalion 17	34 476
SS Pioneer Battalion 17	36 380
SS Flak Battalion 17	46 413
SS Signals Battalion 17	46 104
SS Field Replacement Battalion 17	47 054
SS Supply Services 17	37 306
SS Economic Battalion 17	37 892

Commanding Officers

11/43 – 06/44	SS-Oberführer Ostendorff
06/44 – 10/44	SS-Standartenführer Baum
11/44 – 02/45	SS-Brigadeführer Ostendorff
02/45 – 05/45	SS-Oberführer Bochmann

Wearers of the Knight's Cross of the Iron Cross

23/08/44	SS-Obersturmführer Hinz (Oak Leaves)
23/08/44	SS-Sturmbannführer Wahl (Oak Leaves 01/02/45)
26/11/44	SS-Obersturmführer Kuske
17/12/44	SS-Oberscharführer Gottke
27/12/44	SS-Obersturmführer Papas
19/04/45	SS-Obersturmbannführer Kaiser (Oak Leaves)

18th SS Volunteer Panzer Grenadier Division "Horst Wessel"

(*18. SS-Freiwilligen-Panzergrenadier-Division "Horst Wessel"*)

The 1st SS Motorized Infantry Brigade
(*Die 1. SS-Infanterie-Brigade (mot.)*)

In the course of preparations for the operation in the Soviet Union, on 24 April 1941 the former 8th and 10th SS Death's Head Regiments (*SS-Totenkopf-Standarten*)—now designated the 8th and 10th SS Motorized Infantry Regiments—were combined to form the 1st SS Motorized Brigade. The brigade was initially commanded by *SS-Brigadeführer* Demelhuber, however, he was replaced by *SS-Brigadeführer* Krüger on 25 May 1941. On 21 June the brigade was attached to the *Kommandostab Reichsführer-SS*, and four days later command of the two regiments was transferred to *SS-Brigadeführer* Hermann.

The brigade assembled at SS Training Camp "Debica" (approx. 30 km south of Tarnow) at the beginning of July 1941. There it began training to fight in forests and built-up areas in preparation for combing the occupied areas. On 22 July 1941 the brigade was sent to the Senior SS and Police Commander South, who was based in the Rovno area, approximately 350 kilometers to the east. Three days later elements of the brigade were deployed approximately 40 km to the south, between Ostrog and Shepetovka, to search for enemy stragglers and Jews. Of the latter, about 800 were shot "*for supporting bolshevism.*"

After this first action, the 1st SS Motorized Brigade and the 56th Infantry Division were attached to the commander of Rear Army Area South. On 4 August 1941 elements of the 10th SS Motorized Infantry Regiment shot 1,385 Jewish civilians in the towns of Ostrog, Hrycov, and Kunev. On 29 March 1947 former district court director Otto Albert declared in lieu of an oath:

"While attached to Military Administration Headquarters 787 as judge advocate, in August 1941 our unit was stationed in Ostrog on the then Polish-Russian border.

One day an SS battalion arrived in Ostrog and began dragging all the Jews from their homes and rounding them up.

The military administration headquarters' staff major, Major z.V. Karl Behr, went along and learned that they intended to shoot the approximately 6,000 – 7,000 Jewish civilians outside the town.

Herr Behr was extremely upset by this. He contacted me as judge advocate, and together we immediately drove to where the Jews had been assembled in order to stop the shootings if possible.

At the place we met the commander of the SS regiment, a lieutenant-colonel, with whom we had a very heated argument. Herr Behr pointed out the illegality of the proceedings and declared that it was "plain murder."

The SS commander declared that he was himself extremely uncomfortable about the affair. But he had orders from "the highest level" which he had to obey.

After further negotiation, he finally declared that he was prepared to free all of the Jews employed by the military administration headquarters, whom we had to name, and their families.

When we presented him with a very extensive list, he declared that the entire action would be almost pointless. He nevertheless kept his promise, and many of the unfortunate Jews were able to return to their homes, which had meanwhile been looted by the population. To my knowledge, as a result of our intervention only 400 to 500 Jews were shot that day."

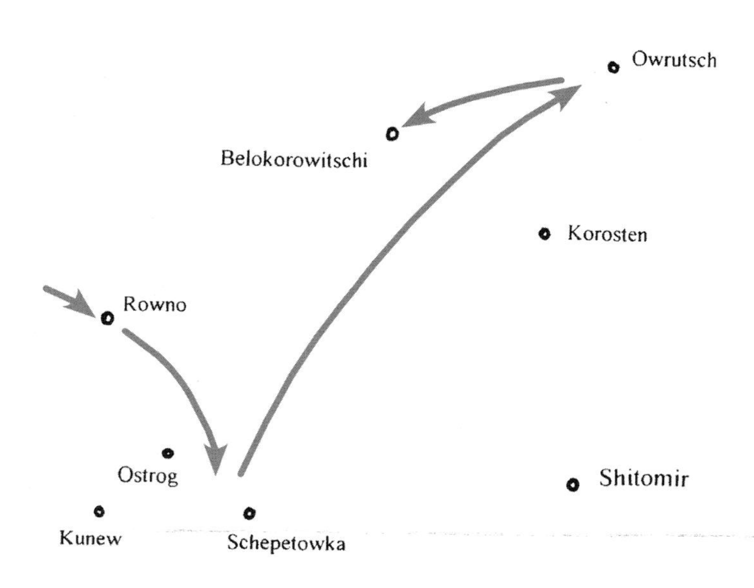

Transfer of the 1st SS Infantry Brigade to SS Training Camp "Debica."

Troops "beautify" a barrack at the training camp by installing a special wooden walkway leading to the entrance.

Two days later the unit moved on to the northeast and combed the area south of Ovruch (rear area of the XVII Army Corps). Approximately 300 Jews were shot in the wooded areas there.

While elements of the 8th SS Infantry Regiment began engaging regular Soviet forces on 20 August 1941, the 10th SS Infantry Regiment was again ordered into action as part of the so-called "ideological struggle." On 24 August, for example, 280 Jews were shot.

The regiment was subsequently ordered to search the area north of the Korosten – Belokorovichi road, and by 12 September 1941 it had shot more than 1,000 Jews and Red Army stragglers.

On 20 September 1941 the unit, since renamed the **1st SS Motorized Infantry Brigade**, moved more than 300 kilometers to the area around Konotop, where its duties included guarding Soviet POWs.

As the brigade was not deployed at the front, at the beginning of December 1941 Himmler planned to send it to the *Leibstandarte SS Adolf Hitler* in the Taganrog area with about 6,000 men and 1,600 vehicles. The Soviet winter offensive in the central sector of the Eastern Front prevented the transfer, however. Instead, on 12 December the brigade set off via Orel for the area south of Verkhovye (approx. 75 km east of Orel), some 400 km away, to join the 2nd Army. As part of XXXIV Army Corps, the brigade was again attached to the 56th Infantry Division. The fighting was fierce at times, and the brigade commander, *SS-Brigadeführer* Hermann, was among those killed. When the Red Army broke through at the Trudy Bend, the 2nd Army made the command of the 10th SS Infantry Regiment deployed there responsible.

When the fighting died down and the units began moving into winter positions, the 1st SS Motorized Infantry Brigade was attached to the 299th Infantry Division within LV Army Corps. The unit remained in this position until 9 August

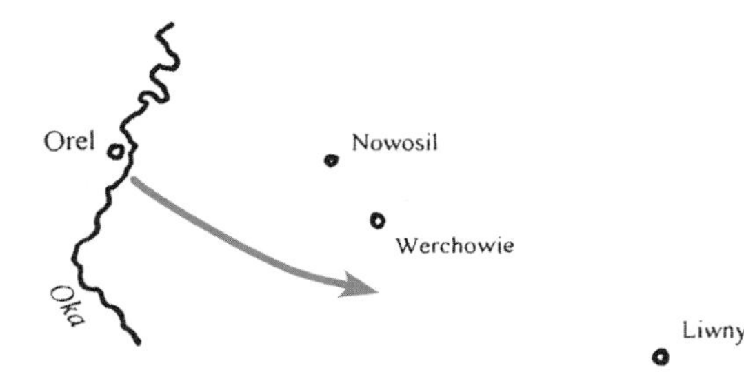

Impressions of Russia…

Members of the 1st SS Infantry Brigade examine a Soviet BT 7 which slid into a water-filled hole.

1942—almost eight months. During this time an exchange of personnel was undertaken. Reich-German personnel were pulled out, and on 26 May 1942 were replaced by 2,300 ethnic Germans from Hungary. At the beginning of August 1942 the Commander-in-Chief of the 2nd Army bade the brigade farewell:

"Originally selected for security duties in the rear areas, called to action at the front in decisive days, the brigade developed into a battle-tested unit."

The brigade was subsequently attached to the Senior SS and Police Commander "Ostland" (*SS-Obergruppenführer* Jeckeln) and moved to the area surrounding Minsk, approximately 700 kilometers away. There it was employed mainly in the rear, largely wooded area against partisans and partisan sympathizers. On 1 September 1942 the units began operating in the Nalibocki Forest (approx. 70 km SW of Minsk). This was followed on 13 September by "Operation Swamp Fever" in the area south of Baranovichi (approx. 130 km SW of Minsk). As in the summer of 1941, this week-long action resulted in the murder of numerous suspected partisans or Jewish civilians. 389 partisans and 1,274 civilians were shot, and 8,350 Jews and 1,217 suspected partisans were deported to labor camps.

The units moved into the Borisov – Orsha area (approx. 100 – 200 km NE of Minsk) for "Operation Karlsbad," which began on 11 October 1942. The operation was an attempt by German and foreign units under the command of *SS-Gruppenführer* von dem Bach to put an end to the almost daily sabotaging of the railway line. More than 1,000 suspected partisans and civilians were shot during the operation.

"Operation Frieda" was a five-day operation in the swamps south of Borisov which began on 5 November 1942. This was followed by "Operation Nuremberg" in the area north of Postavy (approx. 140 km NW of Minsk). Against, casualties of four killed and one slightly wounded, and 715 civilians were shot as suspected partisans.

At the end of the month the 1st SS Motorized Infantry Brigade was alerted and prepared for an action near Velikiye Luki. *Freikorps "Danmark"*[25] was attached to the brigade for the operation. The units arrived in the Gorodok area, 150 kilometers away, on 2 December 1942. From there they were placed under *General der Infanterie* von der Chevallerie within the 3rd Panzer Army to help relieve the vitally important town. Under extreme pressure from the Soviets, however, the town could not be retaken. Instead the German forces, including the 1st SS Motorized Infantry

Russian villages burn during the winter of 1941-42.

Brigade, were forced to withdraw to positions in the rear. On 31 December 1942 the unit reported a strength of:

Officers	Other Ranks	Total
239	5,956	6,135
2.9%	97.1%	100%

When a final relief attempt collapsed under heavy enemy defensive fire, on 16 January 1943 the last German defenders broke out of the town—200 of the original 7,000 men.

Together with the attached *Freikorps "Danmark,"* the 1st SS Motorized Infantry Brigade remained at the front as part of LIX Army Corps until 21 February 1943 and then was temporarily employed in the anti-partisan role. Under the command of the 201st Security Division, from 22 February to 8 March 1943 the units took part in "Operation Ball Lightning," and from 31 March to 2 April 1943 in "Operation Thunderbolt." The objective of both operations was to clear the Vitebsk – Gorodok – Lake Sennitsa rear area.

After briefly manning a sector in the front line, at the beginning of July 1943 the brigade was moved to Borisov (approx. 75 km NE of Minsk). From there the 1st SS Motorized Infantry Brigade took part in "Operation Herrmann" in the Mir – Koidanov – Rakov – Rubiezewicze area (southwest of Minsk) under the Senior SS and Police Commander Russia-Center. The action saw entire regions depopulated. Suspected partisans were deported to labor camps, the cattle were turned over to German agricultural leaders, and the few civilians not under suspicion were placed

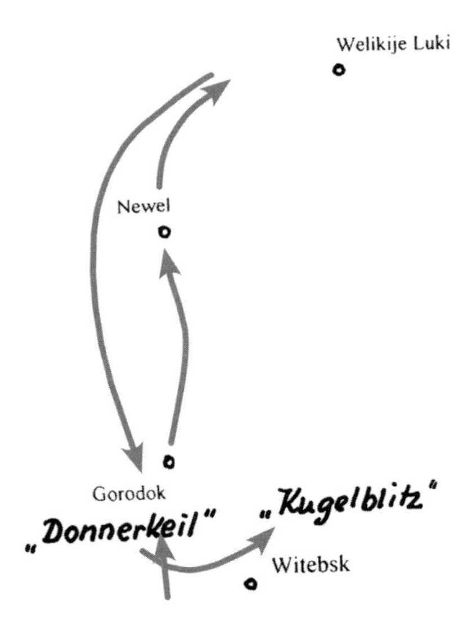

Im Namen des Führers und Obersten Befehlshabers der Wehrmacht

verleihe ich

dem

SS - Sturmmann
Karl C e r m a k,
5. / Inf. - Regt. 8

das

Eiserne Kreuz 2. Klasse

..........Gef. St......, den ..25. Februar..19..42.

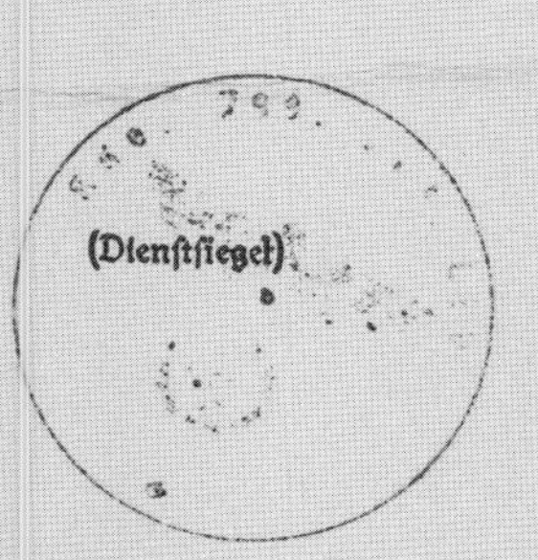

Generalleutnant und Div. Kdr.

(Dienstgrad und Dienststellung)

Generalleutnant Moser, commander of the 299th Infantry Division, to which the 1st SS Motorized Infantry Brigade was attached from the end of December 1941, awarded the Iron Cross, Second Class to SS Sturmmann Cermak.

Besitzzeugnis

Dem SS-Sturmmann
(Dienstgrad)

Karl C e r m a k
(Vor- und Zuname)

5./SS-Inf.-Rgt.8 (mot),
(Truppenteil)

verleihe ich das

Infanterie-Sturmabzeichen

— Bronze —

Im Osten, den 10.März 1942.
(Ort und Datum)

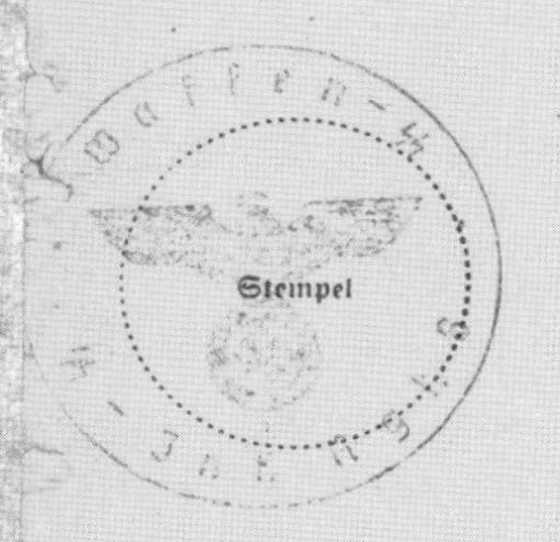

(Unterschrift)

SS-Obersturmbannführer
und Kommandeur
(Dienstgrad und Dienststellung)

The commander of SS Motorized Infantry Regiment 8, SS-Sturmbannführer Tschoppe, awarded SS Sturmmann Cernak the Infantry Assault Badge in Bronze after the difficult winter fighting of 1941-42.

Im Namen des Führers
und Oberſten Befehlshabers
der Wehrmacht

verleihe ich

dem

SS- Sturmmann
Karl C e r m a k
5./SS - J.R. 8

das

Eiſerne Kreuz 1. Klaſſe

Brig.St. Qu. ......, den 2. August 19.42

Der Kommandeur 1.SS-Jnf.-Brigade (mot)

(Dienſtſiegel) SS-Brigadeführer und Generalmajor
der Waffen-SS

(Dienſtgrad und Dienſtſtellung)

On 2/8/42, a short time before his unit was moved to the Borisov area, SS Sturmmann Cermak was awarded the Iron Cross, First Class by *SS-Brigadeführer* Treuenfeld west of Voronezh.

in so-called "closed villages." At that time the 1st SS Motorized Infantry Brigade's reported strength was:

	Officers	NCOs	Enlisted Men	Total
Actual	160	1,015	5,220	6,395
	2.5%	15.9%	81.6%	100%
Combat Strength				
	64	413	2,844	3,321
	1.9%	12.4%	85.7%	100%
Authorized	310	1,512	7,419	9,241
	3.4%	16.4%	80.2%	100%

On 13 August 1943 the Red Army struck at the junction between the 4th Army and the 3rd Panzer Army. At the end of the month the 1st SS Motorized Infantry Brigade was alerted and moved to the Yelnya area (approx. 75 km SE of Smolensk). From there the unit marched via Smolensk into the Yarzevo area (approx. 45 km NE of Smolensk) and was attached to the 25th Panzer Grenadier Division in XXVII Army Corps. On 14 September 1943 there was fierce fighting outside Smolensk. On 24 September the brigade marched through burning Smolensk in the direction of Orsha. To date that month the brigade had suffered losses of 215 killed, 1,173 wounded, and 77 missing. After further heavy fighting in the area west of Smolensk, on 15 October the brigade—by then part of XXXIX Panzer Corps—was named in the Wehrmacht communiqué:

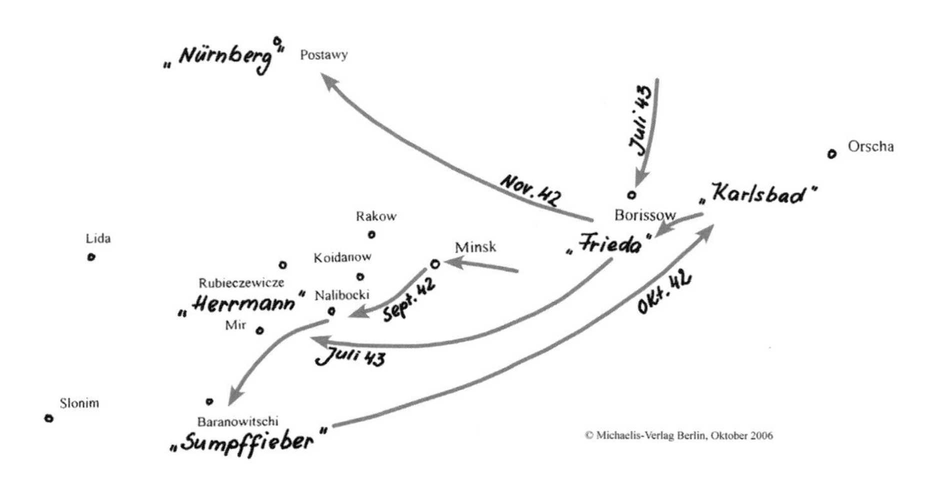

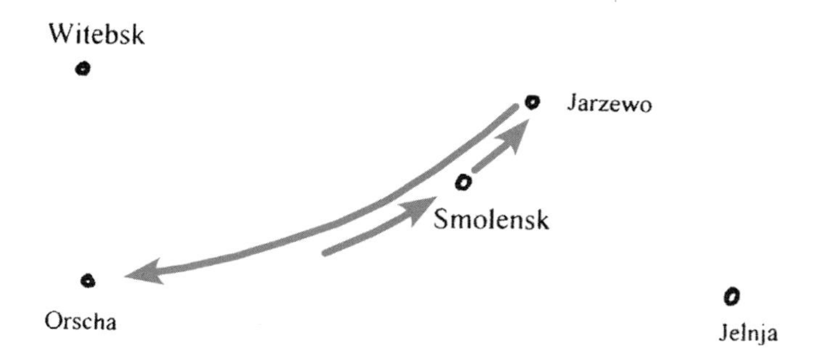

"*Yesterday vigorous breakthrough attempts by the Soviets again failed west of Krichev* (approx. 100 km east of Mogilev, author's note), *and especially west of Smolensk. 46 Soviet tanks were destroyed in the combat zone southwest of Smolensk alone. In the last three days the enemy lost a total of 354 tanks and 233 aircraft in his vain attacks… The 1st SS Motorized Volunteer Grenadier Brigade distinguished itself in the fierce defensive struggle in the central sector of the front…*"

On 22 October 1943 the units of the SS were numbered consecutively. The brigade units received the following designations:

SS Motorized Grenadier Regiment 39
SS Motorized Grenadier Regiment 40
SS Artillery Battalion 51
SS Anti-Tank Company 51
SS Flak Company 51
SS Motorcycle Company 51
SS Signals Company 51
SS Field Replacement Company 51

In mid-November 1943 the brigade had to form a battle group to shore up the hard-pressed front between the Dniepr and Beresina Rivers. The battle group was deployed in front of Bobruisk. After suffering heavy casualties, Battle Group "Wiedemann" was pulled out of action and was assigned to guard the Beresina bridgehead near Parichi. The battle group also fought a number of pitched battles against Soviet partisans.

While the brigade's commander, *SS-Standartenführer* Trabandt, and the bulk of the unit had been withdrawn from action and were on their way to Training Camp Stablack near Königsberg to rest and reequip, SS Battle Group "Wiedemann" was not pulled out of action until the end of December 1943. The battle group then followed the elements of the brigade that had already left.

Officers of the 1st SS Motorized Infantry Brigade.

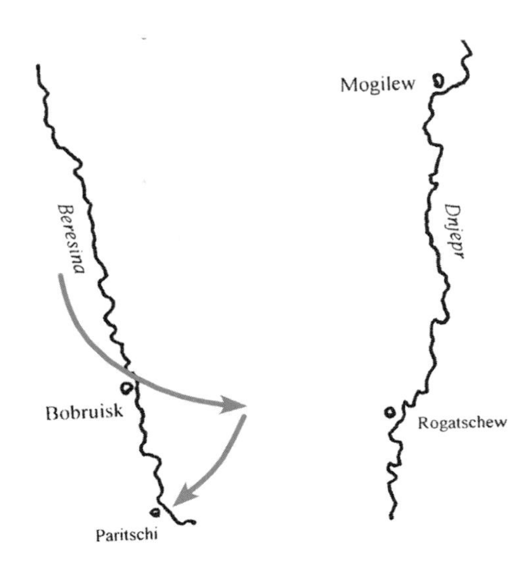

Operations between the Dniepr and the Beresina.

The 18th SS Volunteer Panzer Grenadier Division "Horst Wessel"

The large number of ethnic Germans from Hungary and Rumania called up by the Waffen-SS in late 1943-early 1944 made it possible for it to make good the losses suffered by the existing SS divisions, and also to expand smaller units into divisions. On 25 January 1944 the SS Operational Headquarters announced that the remnants of the 1st SS Motorized Infantry Brigade were to be enlarged into an SS panzer grenadier division. The unit's authorized organization was:

Division Headquarters
SS Panzer Grenadier Regiment 39 (I – III Battalion)
SS Panzer Grenadier Regiment 40 (I – III Battalion)
SS Artillery Regiment 18 (I – III Battalion)
SS Panzer Battalion 18
SS Flak Battalion 18
SS Armored Reconnaissance Battalion 18
SS Anti-Tank Battalion 18[26]
SS Pioneer Battalion 18
SS Field Replacement Battalion 18
SS Panzer Grenadier Training Battalion 18

The division was to be established in the Cilli – Gurkfeld – Rann area of Slovenia while simultaneously providing defense against the local Tito partisans.

On 30 January 1944 Hitler awarded the division the name of the SA-Sturmführer "Horst Wessel."

At the beginning of February 1944 the remnants of the 1st SS Motorized Infantry Brigade began moving from East Prussia to the southeastern theater, where they were attached to LXIX Army Corps, part of the 2nd Panzer Army. With the occupation of Hungary imminent, on 1 March 1944 *Generalleutnant* Foertsch, Chief-of-Staff of Army Group F, ordered the 18th SS Panzer Grenadier Division to form a battle group. The unit's strength was:

	Officers	NCOs	Enlisted Men	Total
Actual	130	938	3,043	4,111
	3.2%	22.8%	74%	100%
Authorized	539	2,877	10,445	13,861
	3.9%	20.7%	75.4%	100%

Its equipment status was:

	Actual	Authorized
Armored vehicles	4	94
Trucks	587	1,580
Automobiles	195	937
Motorcycles	111	806
Heavy anti-tank guns	5	27
Artillery pieces	7	53
MG 42 machine-guns	238	933
8.13-cm mortars	31	99

The available weapons and equipment were not even sufficient for the approximately 4,000 men available to the division, and in the following actions it was able do no more than form battle groups. From 5 to 7 March 1944 units took part in "Operation Melting Snow," which was aimed at partisans in the establishment area.

For the occupation of Hungary SS Armored Reconnaissance Battalion 18 was reinforced. Attached to the 1st Mountain Division, part of LXIX Mountain Corps, it marched out of the area east of Agram to Nagykanisza.

After the many units earmarked as occupation units had to be released to the front, at the beginning of April 1944 the 18th SS Volunteer Panzer Grenadier

Division, which was still in the formation process, was ordered to the Batschka in Hungary.

At the end of April the division was forced to transfer 200 Volkswagens and 300 trucks to the 3rd SS Panzer Division "*Totenkopf*," leaving it almost immobile. On 6 July 1944 *SS-Standartenführer* Trabandt received orders to again form a battle group—this time for employment at the front. The unit, which was about 3,000 men strong, consisted of:

I Battalion, SS Panzer Grenadier Regiment 39
I and II Battalions, SS Panzer Grenadier Regiment 40
I Battalion, SS Artillery Battalion 18
1 company of assault guns from SS Panzer Battalion 18
elements of SS Armored Reconnaissance Battalion 18
elements of SS Signals Battalion 18
elements of SS Medical Battalion 18
elements of SS Flak Battalion 18

The majority of the battle group's soldiers were former members of the 1st SS Motorized Infantry Brigade. By that time the division had a strength of:

	Officers	NCOs	Enlisted Men	Total
Actual	226	1,765	6,539	8,530
	2.7%	20.7%	76.6%	100%
Authorized	539	2,877	10,445	13,861
	3.9%	20.7%	75.4%	100%

The division had thus received barely 4,500 replacements in four months. Some of these had completed basic training, allowing company training to begin.

On 11 July 1944 SS Battle Group "Schäfer" was attached to Army Group North Ukraine, and on 17 July 1944 it relieved elements of the 371st Infantry Division, part of XXIV Panzer Corps, in the area east of Podhajce (Pidhaytsi), on the Strypa River. In the course of the Soviet summer offensive, on 19 July the battle group withdrew to the Koropiec near Podhajce. Drawn into a fighting withdrawal, on 27 July the units reached the Dnestr River west of Khodorov.

At the beginning of August 1944 the SS Battle Group "Schäfer" moved into the Sanok area (approx. 50 km SW of Przemysl) to bolster the 1st Panzer Army's endangered left wing. There it received reinforcements in the shape of a battle group from the SS Waffen Grenadier Brigade "*Charlemagne*."

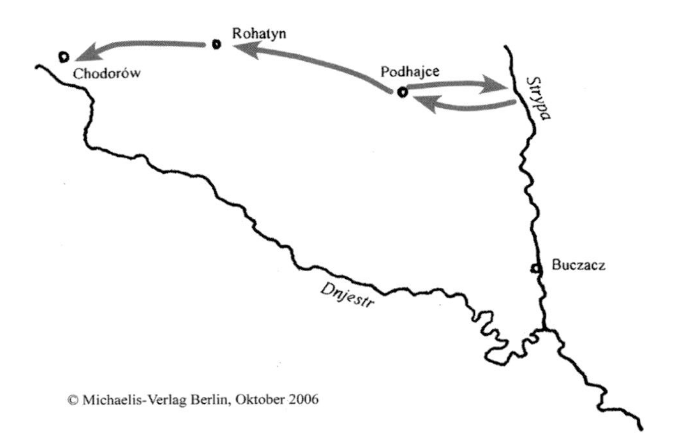

The fighting was fierce and casualties heavy, and initially the German troops were unable to establish contact with the neighboring 17th Army. This only became possible after the front was pulled back by abandoning the much fought-over area.

On 16 August 1944 SS Battle Group "Schäfer" received orders to proceed via Krosno to Radomysl-Wielky (approx. 25 km NE of Tarnow) to LIX Army Corps (17th Army). It subsequently occupied defensive positions south of Mielec on the Wisloka River. On 20 August, following a two-hour artillery barrage, the Red Army took the offensive and the units were forced to pull back. On 20 August 1944 SS Battle Group "Schäfer," by then reduced to two German and one French battalion, was in the area south of Dabrowa (approx. 20 km north of Tarnow). A former member of SS Flak Battalion 18 recalled:

"*In July 1944, before the division was fully established, it was ordered to form a battle group. This included a platoon with three or four guns from our 20-mm flak battery. We had since turned in our four-barreled guns, for by then we were being used in the ground role and they were of no advantage. We moved to Galicia and saw action southeast of Lvov before the battle group was completely assembled. In August 1944 the other platoons arrived with their guns and the battery was again fully functional. The train moved into the area west of Groß-Wardein in eastern Hungary. Our area of operations was from southeast of Ternopol through Podhajce, Stanislau, and Stryj to Sambor. Our 20-mm guns saw plenty of action there and there were fierce battles. Our casualties were considerable, and we had a large number of dead in the fighting.*"

Collar patches worn by members of the 18th SS Volunteer Panzer Grenadier Division "Horst Wessel" (SA Rune).

Wilhelm Trabandt, who had previously commanded the 1st SS Motorized Infantry Brigade in 1943, commanded the 18th SS Division "Horst Wessel" as an SS-Oberführer until January 1945.

Cuff title.

After back and forth fighting, soon afterwards the battle group was pulled out of the main line of resistance and the individual elements were supposed to be sent back to their parent units. For the members of the division, since renamed the **18th SS Volunteer Panzer Grenadier Division**,[28] this meant a transfer to Hungary.

In fact, however, the two remaining battalions were assembled in Skrzyzow, near Tarnow, and on 31 August 1944 they were ordered into eastern Slovakia by Army Group North Ukraine. The Slovakian National Uprising had broken out there and SS Battle Group "Schäfer" was supposed to advance through Käsmark and Deutschendorf (Poprad) into the Rosenberg area. On 6 September 1944 the battle group took Rosenberg and veered toward the southwest.

While the battle group was deployed in Slovakia, in Hungary the establishment of the 18th SS Volunteer Panzer Grenadier Division went on. On 20 September 1944 the division command reported a strength of:

	Officers	NCOs	Enlisted Men	Total
Actual	193	1,209	8,661	10,063
	1.9%	12%	86.1%	100%
Authorized	557	3,388	12,593	16,538
	3.4%	20.5%	76.1%	100%

After the German troops deployed to Slovakia were unable to smash the uprising, and with the Red Army advancing steadily, on 5 October 1944 the *Reichsführer-SS* released further contingents of the Waffen-SS. Among these was the 18th SS Volunteer Panzer Grenadier Division, which was ordered out of the Batschka on the Hungarian-Slovakian border into the Rimavska Sobota – Aggtelek area.

After the battle group again came under the division command, the unit was supposed to begin advancing into central Slovakia on 18 October 1944. SS Volunteer Panzer Grenadier Regiment 40 advanced through Muran toward Tisovec, and there linked up with SS Armored Reconnaissance Battalion 18 coming from Rimavska Sobota. SS Volunteer Panzer Grenadier Regiment 40, which was not yet operational, was held in reserve.

Seen overall, the units met only weak resistance. A former member of SS Armored Reconnaissance Battalion 18 recalled:

"On 24 October 1944 our battalion and assault guns from the division advanced from Tiesovitz (Theißholz) north towards Bresno (Briesen). The attack bogged down toward evening, because the section of track running at right angles to the mountain valley had been blown in the area of the road and a bridge over a small mountain stream when our point company approached. A train with cars loaded with stone drove into the resulting gap, which could not be repaired in the falling darkness.

On the morning of 25 October Wehrmacht pioneers cleared the mines that had been discovered (our pioneer platoon was still being formed in Dresden), and we were able to drive around the obstruction in the riverbed and continue the attack.

The 18th SS Volunteer Panzer Grenadier Division "Horst Wessel" was supposed to be brought up to authorized strength, mainly through the addition of ethnic Germans from Hungary.

In the evening the Slovakians had tried to disrupt our assembly area with their artillery. After they had withdrawn into the mountains lining the road during the day, not only was the battalion able to advance rapidly all the way to Bresno, it also stormed and took the small town and then pressed on through Valeska to Podbrezowa. After one of the assault guns was knocked out and set on fire by a Soviet anti-tank rifle the battalion's riflemen had to carry out the attack on foot. After darkness fell the commanding officer, the commander of the heavy company, and a few men captured a bus full of armed partisans who thought the town was still unoccupied. It proved impossible, however, to take the rest of the town, which extended into a wooded valley. During an attempt to do so the battalion orderly officer was killed and his body could not be recovered until next morning. It was SS-Untersturmführer Molt. He was buried at the edge of the marketplace together with SS-Oberscharführer Gruber, who had been killed during the advance on Breszno when an assault gun suddenly swerved and crushed the Schwimmwagen he was riding in.

On the morning of 26 October 1944 not only had the enemy abandoned Podbrezowa, but they were also withdrawing in panic toward the Tatra. Enemy rearguards offered almost no resistance."

With the Hungarian capital under threat, on 2 November 1944, even before the Slovakian National Uprising was "officially" over, Army Group South asked the German commander in Slovakia for the release of the 18th SS Volunteer Panzer Grenadier Division "*Horst Wessel.*"

Although it had been in formation since January 1944, because of delays in the delivery of weapons and equipment and bottlenecks in training ten months later the division was still far from being a frontline unit. It was therefore attached to the 4th SS Police Panzer Grenadier Division within LVII Panzer Corps (6th Army) and occupied defensive positions in the Jasz-Ladany area (approx. 20 km north of Szolnok). When the Red Army took the offensive on 9 November 1944, the result was a disaster for the deployed elements of the 18th SS Volunteer Panzer Grenadier Division. On that day Army Group South's war diary recorded:

"*In the sector held by the 4th SS Polizei Panzer Grenadier Division, part of LVII Panzer Corps, elements of the 18th SS Division failed completely, allowing themselves to be overrun and surrendering.*"

The reason behind this failure by the ethnic Germans, most of whom had been forced into the service, was inadequacies in command, training, and armaments. A former member recalled:

"Creation of the new 18th SS Panzer Grenadier Division "Horst Wessel" was supposed to take place in the Obertschernowitz – Radotin area. The enlisted men were ethnic Germans recruited from the Hungarian provinces of Baranya and Tolna and assembled at Training Camp "Püspökladány" (eastern Hungary in the Komitat of Hajdú-Bihar). In the first half of October, after a long train ride, about 1,300 men arrived to form the pioneer units. It must be said that 200 men deserted on the way there.

As most of the men had no pre-military training and only a few had volunteered for military service, a training period of 16 weeks was decided on, an extremely long time compared to the training times for draftees from the Reich. The shortage of instructors also made itself felt and the unit leaders were very much overworked. Nevertheless, on 11 November 1944 the reinforced battalion stood in front of the city hall in Prague to take the oath before the Reich Protector for Bohemia-Moravia, Karl-Hermann Frank. Organized by SS-Hauptsturmführer Fink, this big event, with participants from other units and organizations, may have been the last major public event held by the Germans in Prague.

The tough training, the first lengthy separation from family for most of the soldiers, and the infrequency of mail from their home villages placed a great mental strain on the men, especially as some were already fathers."

By 13 November 1944 the units had been driven back into the Jasz-Bereny area. Two days later they occupied the so-called "Karola Position" to defend Budapest. There was fighting near Aszod, and by 25 November 1944 the division had retreated into the Ecseg area, about 30 kilometers to the north.

When the Red Army launched its operation against Budapest on 5 December 1944, the 18th SS Volunteer Panzer Grenadier Division "*Horst Wessel*" and the 4th SS Polizei Panzer Grenadier Division came under the command of IV Panzer Corps of the 8th Army. When the heavy fighting began some units again left their positions and most men threw their weapons away.

Beginning on 16 December 1944, the division was pulled out of action to be reorganized in the Schemnitz – Karpfen area in Slovakia. At the beginning of January 1945 it was ordered into the Marburg – Cilli area to be reformed.

The powerful I Battalion, 18th SS Volunteer Panzer Grenadier Regiment 40 did not accompany the division; instead, together with its many heavy weapons, it was incorporated into the 4th SS Polizei Panzer Grenadier Division.[29] On 13 January 1945 the Wehrmacht communiqué declared of this battalion:

"In seven days of defensive fighting south of the Slovakian border, SS-Sturmbannführer Riepe, a member of an SS volunteer panzer grenadier division [...], repulsed every attack by the superior enemy, inflicting heavy losses, and courageously held the town of Czesani (approx. 16 km east of Balassa-Gyarmat) as a pillar of the German front."

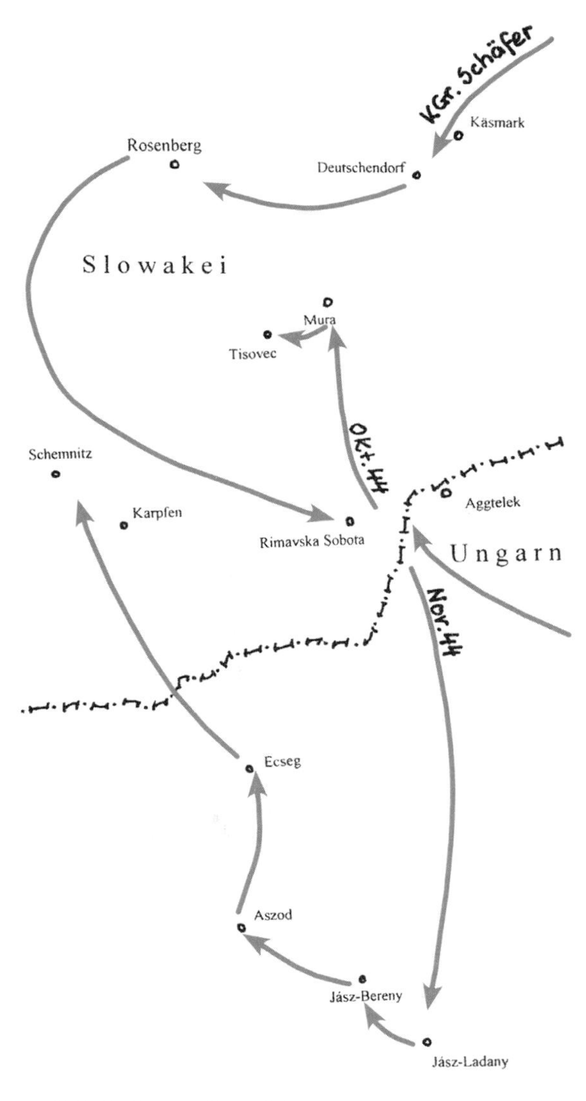

In Lower Styria, the 18th SS Volunteer Panzer Grenadier Division "*Horst Wessel*" was to be issued new equipment, and its fighting strength heightened through the addition of Reich German personnel. On 18 January 1945 the headquarters of Army Group South wrote of the unit:

"*Combat value IV. Consists mainly of ethnic Germans from Hungary. Not up to crisis situations. Still only remnant units on hand. Command not strict enough. The Reichsführer-SS has already ordered its officers replaced.*"

In the course of this replacement of officers, *SS-Standartenführer* Bochmann assumed command of the remnants of the division. The ratio of Reich to ethnic Germans was to be equalized through the incorporation of Luftwaffe personnel. As the former air force personnel—some of whom were to take over lower and middle command positions—brought almost no infantry combat experience with them, the desired increase in fighting strength remained no more than a theory.

Just two weeks later, because of developments in the situation in Silesia, the "division" was ordered out of the Marburg area to Mährisch-Ostrau, 440 km to the north. Attached to XI Army Corps, part of the 1st Panzer Army, in the first weeks of February the regiments occupied positions south of the Soviet Oder bridgehead near Cosel. At that time the unit had:

31 assault guns
12 light field howitzers
2 heavy field howitzers
285 light machine-guns
83 heavy machine-guns
800 trucks
359 passenger cars
204 motorcycles
43 prime movers

This strength was equivalent to that of an artillery battalion or a weak panzer battalion equipped with assault guns. On 16 February 1945 the Red Army launched an attack from the Cosel bridgehead. In the fighting that lasted until 23 February, the 18th SS Volunteer Panzer Grenadier Division successfully held its positions. On 8 March 1945 the XI Army Corps ordered an attack against the enemy bridgehead. Elements of the 18th SS Volunteer Panzer Grenadier Division took part in the unsuccessful operation. One week later the Red Army launched a major offensive, shattering the entire defensive front in Silesia. Soviet troops also succeeded in encircling large elements of *Korpsgruppe Silesia* (*Korpsgruppe* = corps-size ad hoc formation) and XI Army Corps in the area west of Oppeln to Cosel. A former member recalled:

"The moment came on 8 February 1945. We entrained in Radotin and headed for Upper Silesia. We detrained in Mährisch-Ostrau. My pioneer company, which was attached to SS Panzer Grenadier Regiment 40, marched via Petershoven to Koblau, and there we again met elements of SS Pioneer Battalion 18. We were ordered to undertake an operation from there and twelve trucks were supposed to come for us. Only eight were available, however, and for the time being the dream of a motorized unit was replaced by the dream of a horse-drawn one, but even this did not come true, for the principal mode of transport and conveyance was our own two legs. Elements of the regimental headquarters were quartered in Gnadenfeld, which is where the pioneer company was sent. From there we marched on foot to Groß-Neukirch, where we were to see action. On 16 February 1945 the Russians attacked the town with heavy artillery fire, which inflicted the first serious casualties on the company. Uncertainty as to the position of the front and individual massed assaults by the Russians kept us constantly on the move. The company had just settled into the Schwerfelde estate when Russian men, horses, and vehicles began entering the estate from the road. The company was able to withdraw silently and then lay down accurate fire on the estate with two mortars. A night patrol entered Groß-Neukirch and moved about among the Russians without being spotted. It returned with several items the company had left behind there. With the ground frozen solid most infantry actions resulted in heavy casualties, as the Russians were outstanding snipers.

After a daylight attack on the village of Klein-Ellgut had failed at the cost of 80 dead, the company received orders to retake the village in a night attack by a pioneer assault team. Exploiting the terrain, the team went around the Russians and attacked from the east. Surprise was total, and the 27 pioneers retook the village with hand grenades and panzerfausts. Our losses: 1 dead and 4 wounded.

The losses suffered by SS Pioneer Battalion 18 and the other pioneer units of the 18th SS Panzer Grenadier Division "Horst Wessel" were extremely heavy, for after four weeks of action personnel strength was only a fraction of what it had been at the start. The Russians had moved large forces into their bridgehead on this side of the Oder and committed their spearheads, supported by heavy weapons, relatively skillfully, making it impossible for us to establish a continuous front line. The German troops had almost no heavy weapons or tanks, and there were frequent shortages of fuel and ammunition. Some of our losses were attributable to a not inconsiderable number of desertions. The Russian front-line propaganda made great use of these people, naming them in loudspeaker broadcasts and leaflets and declaring that they were already on their way home."

On 19 March 1945 the surrounded German units launched a breakout to the west. While the last assault guns of SS Panzer Battalion 18 were supposed to spearhead the breakout, the two panzer grenadier regiments were to cover the breakout to the east. On the morning of the next day the soldiers reached the German lines near Hotzenplotz. The breakout had succeeded at a heavy cost in personnel and equipment.

By 21 March 1945 the remnants of the division had assembled in the Obersdorf area. Small battle groups were sent into action from there several times, until on 26 March the unit was ordered to move to the Karlsbrunn area (approx. 28 km south of Ziegenhals). Orders were issued for the reorganization of the division, whose strength at that time was:

	Officers	NCOs	Enlisted Men	Total
Actual	150	1,052	4,010	5,212
	2.9%	20.2%	76.9%	100%
Authorized	571	3,406	12,930	16,907
	3.4%	20.2%	76.4%	100%

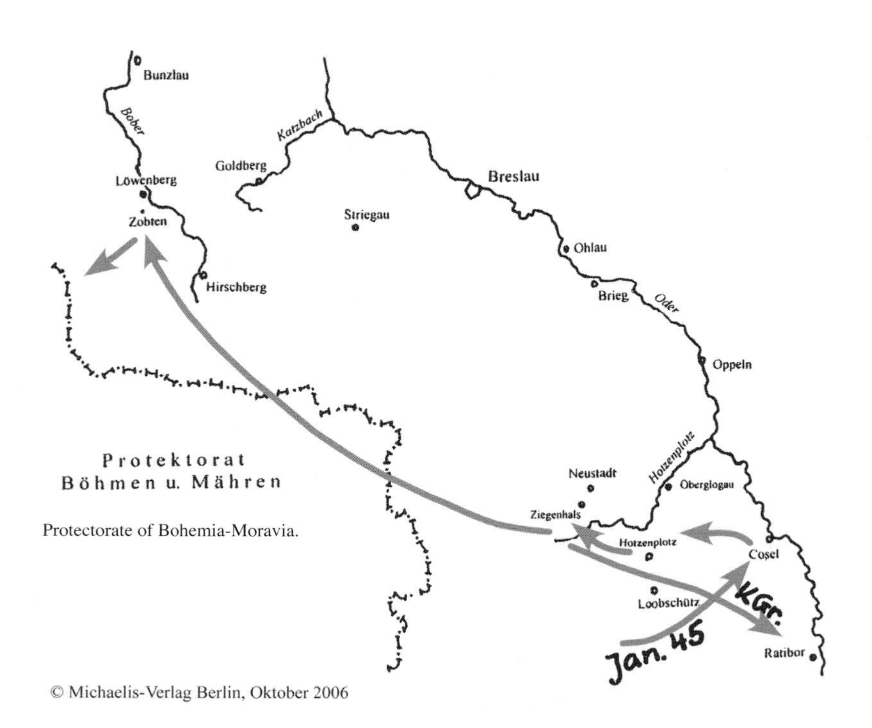

Protectorate of Bohemia-Moravia.

While I and II Battalions, SS Volunteer Panzer Grenadier Regiment 39, and I and II Battalions, SS Volunteer Artillery Regiment 18 were supposed to be newly formed at the SS Training Camp "Bohemia and Moravia," the remaining battalions and forces from the supporting units functioning as infantry were to be combined to form a regiment-strength battle group. Under the command of *SS-Sturmbannführer* Schumacher, the battle group reached the Ratibor area and was attached to XI Army Corps. The unit was shattered in at times heavy fighting in Ratibor and the surrounding area. Finally it was attached to the 371st Infantry Division and was captured by the Soviets.

On 10 April 1945 the last elements of the 18th SS Volunteer Panzer Grenadier Division were ordered to the front with a total of:

6 Sturmgeschütz III
7 self-propelled 75-mm anti-tank guns
1 towed 88-mm anti-tank gun

Elements of the two SS panzer grenadier regiments and SS Armored Reconnaissance Battalion 18 subsequently occupied positions within VIII Army Corps in the Zobten area on the Bober River (approx. 6 km SE of Löwenberg). As the main effort was on the right wing of the Soviet army group there was no further heavy fighting in the Hirschberg area before the surrender.

On 7 May 1945 the division commander, *SS-Standartenführer* Petersen, received orders from VIII Army Corps to disengage from the enemy and march to the demarcation line. Initially the units moved in an orderly fashion into the Reichenberg area and from there tried to reach the Elbe in small groups. Many members of the division were killed by Czech partisans during the attempt. A former member recalled:

"*In January 1945 I joined SS Pioneer Battalion 18 as a clerk in Mährisch-Ostrau. As far as I could ascertain, most of the functionaries in the entire battalion were from other SS divisions, while the German-speaking enlisted men were Hungarian. The fighting in Schwarzwasser was probably the first for these people. I heard that they laid down their weapons. In any event, the entire battalion had to form up and the division judge appeared with SS-Pionier Scherzel. He had him shot in front of us, for cowardice in the face of the enemy. He threatened everyone. Then we had to sing a marching song. The morale in the unit afterwards was catastrophic! The enlisted men only spoke to each other in Hungarian. They were convinced that their homeland was as good as lost. Whether or not they had been forced into the Waffen-SS I do not know. We retreated through Ratibor into the Altvater Mountains. Final order: assemble in Gablonz. On 9 May 1945 I met SS-Untersturmführer Berger, who stood on the cupola of a tank and called out:*

Krause, you fool, get out of here! That was a clear indication. It wasn't until we were just outside Gablonz that a Wehrmacht Oberleutnant told us that Germany had surrendered. The drivers set their vehicles on fire and we destroyed our SS badges and put on our fatigues. That saved us from a painful, unworthy death at the hands of the Czechs. Everything I experienced from then on was terrible. In Turnau I and two friends joined a group of Wehrmacht personnel. We watched as, about 100 meters away, Czechs searched the prisoners of war for SS. They were then worked over with sabers and the like and shot in a clay pit. After the Waffen-SS had been sorted out the Russians took charge of us. They took us to Neuhammer/Queiss and our lives as prisoners of war began."

Planned as an SS panzer grenadier division, this unit never attained its full strength and employability. Despite ten months of training, its first action at the front turned into a fiasco. Shortcomings in leadership and training and the absence of heavy weapons kept the division from becoming operational. Bottlenecks in the German armaments industry also meant that, even with a complement of 700 trucks, the unit, which was designated a panzer grenadier division, was still short about 900. Consequently the division was limited to operations by regiment-strength battle groups.

Military Postal Numbers	
Division Headquarters	16 441
SS Volunteer Panzer Grenadier Regiment 39	18 830
I Battalion	19 087
II Battalion	19 664
III Battalion	20 387
SS Volunteer Panzer Grenadier Regiment 40	22 110
I Battalion	22 782
II Battalion	22 948
III Battalion	23 331
SS Volunteer Artillery Regiment 18	01 026
I Battalion	56 218
II Battalion	07 298
III Battalion	03 427
IV Battalion	06 367
SS Panzer Battalion 18	57 906
SS Repair Battalion 18	06 485
SS Flak Battalion 18	07 579
SS Signals Battalion 18	18 610
SS Armored Reconnaissance Battalion 18	18 610
SS Anti-Tank Battalion 18	07 947

SS Pioneer Battalion 18	01 165
SS Supply Services 18	06 485
SS Economic Battalion 18	08 705
SS Field Replacement Battalion 18	05 694
SS Panzer Grenadier Training Battalion 18	64 232

Commanding Officers

04/41 – 06/41	*SS-Brigadeführer* Demelhuber
06/41 – 12/41	*SS-Brigadeführer* Richard Herrmann
01/42 – 07/42	*SS-Brigadeführer* Hartenstein
12/43 – 12/44	SS-Oberführer Trabandt
01/45 – 03/45	SS-Standartenführer Bochmann
03/45 – 05/45	SS-Standartenführer Petersen

Wearers of the Knight's Cross of the Iron Cross

14/10/43	SS-Standartenführer Schäfer
13/11/43	SS-Standartenführer Petersen
01/12/43	SS-Sturmbannführer Hörnicke
10/12/43	SS-Obersturmführer Sonne
27/12/43	SS-Obersturmführer Rubatscher
06/01/44	SS-Oberführer Trabandt
02/01/45	SS-Hauptsturmführer Dr. Lipinski
13/01/45	SS-Sturmbannführer Riepe
30/03/45	SS-Standartenführer Bochmann (Swords)

23rd SS Volunteer Panzer Grenadier Division "Netherland" (Dutch No. 1)

(*23. SS-Freiwilligen-Panzergrenadier Division "Nederland" (niederländische Nr. 1)*)

The Volunteer Legion "Netherlands"
(*Die Freiwilligen-Legion "Niederlande"*)

After the end of the Western Campaign, the SS Headquarters in the Netherlands began recruiting volunteers for the SS Regiment "*Westland*" formed in August 1940. As of 10 April 1941, Dutch volunteers who were unable to meet the high recruitment standards could be assigned to the **SS Volunteer Regiment "Northwest**."

In November 1941, for propaganda reasons, orders were issued for the formation of national legions to take part in the "struggle against Bolshevism." The

The "Wolfsangel" collar patch worn by members of the Legion and the later 23rd SS Volunteer Panzer Grenadier Division "Nederland."

The nationality badge in the colors red – white – blue.

Cuff title.

SS Volunteer Regiment "Northwest" was subsequently disbanded at Training Camp Arys in East Prussia and the Volunteer Legions "Netherlands" and "Flanders" were established. On 9 January 1942 the Volunteer Legion "Netherlands," consisting of three battalions plus a 13th and 14th Company, reported a strength of:

Officers	NCOs	Enlisted Men	Total
67	331	2,536	2,934
2.2%	11.3%	86.5%	100%

Dutch numbers were:

26	2	2,179	2,207
1.2%	0.1%	98.7%	100%

Only 28 of 398 officers and NCOs were Dutch, and this, despite Himmler's order that the legions should be led by fellow countrymen.

A few days later the unit was transferred to the front north of Selo Gora (approx. 40 km NW of Novgorod) and attached to the 20th Motorized Infantry Division within XXXVIII Army Corps. The bulk of the regiment's personnel had been with the unit or its predecessor for six months or longer, and some had served in the Netherlands Army. The Volunteer Legion "Netherlands" was thus considered battle ready.

In the course of their winter offensive that began on 13 January 1942, at the end of the month the Red Army also attacked the Dutch positions near Selo Gora – Gusi. There followed fierce, changeable fighting, in which contact between units was sometimes broken. Only with great effort was a new main line of resistance created. A former member recalled the unit's first action:

"As an SS-Sturmmann I was transferred to the Volunteer Legion "Netherlands" at Training Camp Arys, where I was assigned to the 11th Company. Altogether there were 15 Reich Germans serving in the company—I was a squad leader. After basic and combat training, in January 1942 we were transported by rail to Danzig. III Battalion was loaded onto the transport "Levente" and sailed to Libau. In the port there were sleds and toboggans which we added to our equipment. Two toboggans were lashed together and loaded with machine-gun boxes, baggage, and machine-guns. Gunners 1 and 2 pulled, Gunner 3 and the squad leader pushed. Other squads and platoons did the same. A gray party snaked its way northwards through the snow-covered landscape.

After exhausting marches in bitter cold we reached Pskov. Artillery fire could be heard in the distance. We encountered the first wounded Spaniards. We were quartered in Selo Gora and given a few days of well-earned rest. Snow smocks were issued. The 11th Company was given a mission, to guard the Gusi – Radoni road. Small bunkers had been built every 200 meters along the road. Four men were assigned to each bunker. In the awful cold partisans scurried through the lines. A few bunkers were ambushed. Every man was warned not to fall asleep. Once a day, at about eight in the morning, the senior NCO brought the rations. It was rumored that I Battalion was encircled and that we were supposed to attack and break them out.

On the night of 27 February 1942 we moved into the assembly area. We lay in the snow—cold, very cold—but nevertheless no one froze anything. The next day the platoons and squads formed up for the attack. It began with mortar fire. Through deep snow, ducking behind small trees, we made rapid progress at first. I saw the first fallen comrades. At about 10:00 hours I felt a blow on my upper right arm, and then two men from my squad dragged me to the rear. This wound ended my action with the Legion."

In the fighting in the snow-covered landscape, the Volunteer Legion "Netherlands" suffered the following losses by 23 March 1942:

	Officers	NCOs	Enlisted men
Killed	4	74	78
Wounded	11	428	439

After the end of the first fighting at the so-called "Volkhov pocket" the front in the Selo Gora area died down. On 14 April 1942 the 126th Infantry Division took over the sector of front and with it tactical command of the legion. When the 2nd SS Motorized Infantry Brigade moved into the main line of resistance the Legion was attached to it.

On 7 July 1942 the Legion launched a successful attack from its positions toward the Novgorod – Leningrad rail line. New positions were immediately constructed before Selo Gora. On 20 June 1942 the battalions took the offensive in the swampy wooded terrain and captured Maloye Zamoshye. By 29 June 1942 the Legion had taken about 3,500 prisoners and captured 380 trucks, 66 guns, 210 machine-guns, and 50 mortars. The next day the unit was pulled out of action and ordered to Selo Gora to rest and reequip.

On 21 July 1941 the Volunteer Legion "Flanders" was attached to the Volunteer Legion "Netherlands." Designated Battle Group Fitzhum, the units were transferred to L Army Corps outside Leningrad the next day. At the beginning of August 1942 the last elements arrived by rail in the new area of operations near Krasnoye Selo (approx. 20 km south of Leningrad).

A group of Dutch volunteers in Vaught.

An SS volunteer on home leave. He has already seen action, as evidenced by the Iron Cross, Second Class, the East Medal, the Wound Badge and the Infantry Assault Badge.

Employed only in the siege of Leningrad, the unit saw no further serious action. On 31 December 1942 the Volunteer Legion "Netherlands" submitted the following strength report:

Officers	Other Ranks	Total
57	1,698	1,755
3.3%	96.7%	100%

Having received few replacements from the Netherlands, the Legion's personnel strength had fallen to that of a weak regiment. On 9 February 1943 the 1st Company was awarded the honorary title "General Seyffardt." Three days earlier the general had been shot by the Dutch resistance movement for collaborating with the Germans.

The next day the Red Army launched an attack from Kolpino. Stalled by a fierce German defense, the Russians shifted their main effort to Krasny Bor. The fighting was fierce and casualties heavy, but the Volunteer Legion "Netherlands" held its positions.

On 27 April 1943 the unit was withdrawn from the 2nd SS Motorized Infantry Brigade, which until then had consisted of Battle Group Fitzhum and three Latvian *Schutzmannschafts* (Police) Battalions. With the removal of the Volunteer Legion "Netherlands," the brigade received three new Latvian police battalions and became the Latvian SS Volunteer Brigade.

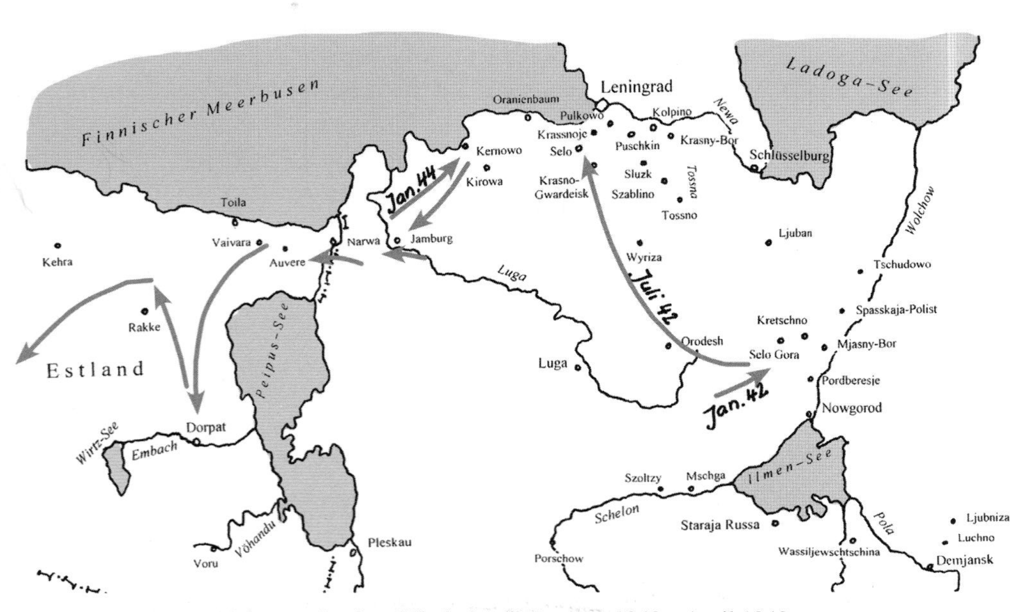

Volunteer Legion "Niederland": January 1942 – April 1943.
SS Volunteer Panzer Grenadier Brigade: January – September 1944.

With the recruitment of large numbers of ethnic Germans from Rumania in the spring of 1943, a program began to enlarge smaller units. In order to reap propaganda benefits, there were also plans to form a "Dutch" division with the help of ethnic Germans. Orders for establishment of the division were issued on 19 July 1943, however, on 23 October the unit's size was reduced to that of a brigade. With only about 2,200 Dutch volunteers, Himmler realized that it would be impossible to form a "Dutch" SS division with almost 14,000 men.

The 4th SS Volunteer Panzer Grenadier Brigade "Netherland" (*Die 4. SS-Freiwilligen-Panzergrenadier-Brigade "Nederland"*)

Initially transferred from the Eastern Front to Training Camp "Mielau," from there the Dutch unit was ordered into the Sonneberg – Hildburghausen – Schleusingen area of Thuringia. In summer 1943 it was moved to the Balkans to strengthen the German forces there in preparation for the predictable Italian departure from the Axis alliance.

The rail transports arrived in the area north of the Save River at the end of August – beginning of September 1943. There the troops were to guard the area against partisans while at the same time completing establishment of the brigade. The planned organization of the brigade looked like this at the end of October 1943:

Brigade Headquarters	Headquarters Krapinski
SS Volunteer Panzer Grenadier Regiment 48 "General Seyffardt" (Dutch No. 1)[30]	Zabok
I Battalion	Krapina
II Battalion	Zabok
SS Volunteer Panzer Grenadier Regiment 49 "de Ruyter" (Dutch No. 2)[31]	Stubicatoplice
I Battalion	Donja-Stubica
II Battalion	Oroslawje
SS Volunteer Artillery Regiment 54	
I Battalion	
II Battalion[32]	
SS Pioneer Battalion 54	
SS Anti-Tank Battalion 54	
SS Reconnaissance Company 54	
SS Medical Company 54	
SS Signals Company 54	
SS Economic Company 54	
SS Supply Services 54	

Jürgen Wagner—seen here as an SS-Oberführer—was given command of the SS Volunteer Panzer Grenadier Brigade "Nederland" in the summer of 1943. He received the German Cross in Gold on 8/12/42, the Knight's Cross of the Iron Cross on 24/7/43, and the Knight's Cross with Oak Leaves on 11/12/44.

During the establishment period in Croatia there were frequent operations against Tito partisans. As a result the brigade commander, *SS-Brigadeführer* Wagner, was hanged in Belgrade on 5 April 1947. George Duiker, a Dutchman who rose to become an *SS-Untersturmführer*, recalled:

"I was born in Zandvoort on 4 November 1922. When I reported to the labor service in August 1940 they wanted to send me to Germany to work in a coal mine. As I didn't want that, and probably because of my size, the official wearing the party badge of the NSDAP in Holland suggested that I join a "sports club" in Germany! As I preferred that to working in a coal mine I accepted, and in September 1940 found myself in the "Munich-Freimann" SS barracks, specifically in the 9th Company of the SS Regiment "Westland." And I did play a lot of sports there.

On 14 August 1941 near Kremenchug I was wounded by an explosive bullet, and at the beginning of 1942 I rejoined the 6th Company, "Westland" at the front. I was promoted to SS-Unterscharführer at the end of October 1942 and ordered to attend the first course for Germanic officers at the SS officer's school in Tölz. After passing the test I was granted leave, after which I was supposed to report to Training Camp "Grafenwöhr." My former division commander Steiner was then forming the III (Germanic) SS Panzer Corps there. Actually I wanted to return to my old company in the SS Regiment "Westland," but being a Dutchman I was assigned to the new 4th SS Volunteer Panzer Grenadier Brigade "Netherland."

I was able to join the SS Assault Gun Battery. As we had not been issued assault guns, we had 75-mm anti-tank guns with nothing to tow them—the soldiers therefore had to pull the things themselves! After a burst barrel we were down to two guns. There was an alert and we were loaded onto trains. No one knew where we were going. Then we learned that we were heading south. We travelled to Croatia. There we were to finally complete the establishment of the brigade, disarm the Italians, and fight the partisans.

The latter were regularly carrying out attacks, and we were constantly employed to guard railway bridges. By day the partisans pretended to be normal civilians and worked in the fields.

At first my commander was an older SS-Sturmbannführer from North Schleswig. As he was not enthusiastic enough, he was replaced by a younger SS-Obersturmführer from the Sudetenland. Soon afterwards he gave our rations away to a "Strength through Joy" dance group. He, too, was replaced, and we got a Wehrmacht Hauptmann who had switched to the Waffen-SS. He had previously been stationed in Norway. When it was announced that we were being transferred to the Eastern Front he issued himself a leave pass and disappeared home to the Saarland. His adjutant contracted a venereal disease in Croatia and had to be

Soldbuch
zugleich Personalausweis
Nr.

für
(Dienstgrad)
1.7.42 (Datum) SS-Sturmmann
1.3.1943 (neuer Dienstgrad) SS-Unterscharführer
Henk Ophoff
(Vor- und Zuname)
Beschriftung und Nummer der Erkennungsmarke
Blutgruppe Blutgruppe 0 (Null)
Gasmaskengröße 2
Wehrnummer

geb. am 14-5-16 in Amsterdam
(Ort, Kreis, Verw.-Bezirk)
Religion R.K. Stand, Beruf verkaufer

Personalbeschreibung:

Größe 1.86 Gestalt schlank
Gesicht oval Haar dunkel blond
Bart — Augen grau
Besondere Kennzeichen (z. B. Brillenträger):
Brillenträger
Schuhzeuglänge Schuhzeugweite
Henk Ophoff
(Vor- u. Zuname, eigenhändige Unterschrift des Inhabers)

Die Richtigkeit der nicht umrandeten Angaben auf Seite 1 und 2 und der eigenhändigen Unterschrift des Inhabers bescheinigt

den 26 April 1941
(Ausfertigender Truppenteil, Dienststelle)
SS-Hauptsturmführer und Kompaniechef
(Eigenh. Unterschrift, Dienststellg. d. Vorgef.)

2

Bescheinigungen
über die Richtigkeit der Zusätze und Berichtigungen auf Seiten 1 und 2

Lfd. Nr.	Art der Änderung	auf Seite	Datum	Truppenteil	Unterschrift	Dienstgrad und Dienststellung
1.		4				SS-Hauptst.
2.	Beförderung	1	1.3.43	Stabskomp.		
3						

3

Dutch volunteer Henk Ophoff joined the replacement battalion of the SS Regiment "Westland" on 26/4/41, and was initially assigned to the Field Replacement Battalion of the SS Division "Wiking." When the 4th SS Volunteer Panzer Grenadier Brigade "Nederland" was formed he was transferred to SS Reconnaissance Battalion 54.

A. Zuletzt zuständige Wehrersatzdienststelle:

B. Zum Feldheer abgesandt von:[1])

	Ersatztruppenteil	Kompanie	Nr. der Truppenstammrolle
a	Stammkomp./Gen.-Ausb.-	45/1692	
b	u. Ers.-Batl. der Waffen-SS		
c			

C.

	Feldtruppenteil[2])	Kompanie	Nr. der Kriegsstammrolle
a			
b			458
c			

D.

Jetzt zuständiger Ersatztruppenteil[2])	Standort
~~Stabskomp. d. Waffen-~~	~~Frankfurt a. M~~
~~b. Röntgensturmbann~~	
~~Ers. Batl. d. Waffen-SS~~	~~Stettin~~

(Meldung dortselbst nach Rückkehr vom Feldheer oder Lazarett, zuständig für Ersatz an Bekleidung und Ausrüstung)

[1]) Vom Ersatztruppenteil einzutragen, von dem der Soldbuchinhaber zum Feldheer abgesandt wird.

[2]) Vom Feldtruppenteil einzutragen und bei Versetzungen von einem zum anderen Feldtruppenteil derart abzuändern, daß die alten Angaben nur durchstrichen werden, also leserlich bleiben.

Weiterer Raum für Eintragungen auf Seite

4

Anschriften der nächsten lebenden Angehörigen

des .... (Vor- und Zuname)

1. Ehefrau: Vor- und Mädchenname

(ggf. Vermerk „ledig")

Wohnort (Kreis)

Straße, Haus-Nr.

2. Eltern: des Vaters, Vor- und Zuname

Theodor Ophoff

Stand oder Gewerbe

der Mutter, Vor- u. Mädchenname

Wohnort (Kreis)

Straße, Haus-Nr.

3. Verwandte oder Braut:*)

Vor- und Zuname

~~Stand oder Gewerbe~~

Wohnort (Kreis)

Straße, Haus-Nr.

*) Ausfüllung nur, wenn weder 1. noch 2. ausgefüllt sind.

5

George Duiker was also a member of the SS Panzer Grenadier Brigade "Nederland" from the summer of 1943.

hospitalized. Our doctor—the Dutch Dr. Lessing—committed suicide after it was learned that he was a homosexual who had repeatedly abused recruits."

On 21 November 1943 Hitler had made it clear that the III (Germanic) SS Panzer Corps was to be transferred to the Eastern Front along with the 11th SS Volunteer Panzer Grenadier Division "*Nordland*" and the 4th SS Volunteer Panzer Grenadier Brigade "Netherland." The units began leaving for the east in mid-December. At the end of December 1943 the brigade reported a strength of:

	Officers	NCOs	Enlisted Men	Total
Reich Germans	94	497	457	1,044
Ethnic Germans	1	32	2,115	2,148
Dutch	33	199	1,984	2,216
Other	1	--	13	14
Total	129	728	4,569	5,426
	2.4%	13.4%	84.2%	100%

These figures show that the unit had attained at least 80% of its authorized strength. After only about 2,200 Dutchmen and approximately 1,000 Reich Germans (core personnel) were available, an initial group of almost 2,200 ethnic Germans from Rumania were assigned to the 4th SS Volunteer Panzer Grenadier Brigade "Netherland."[33] There were difficulties associated with the delivery of heavy weapons and equipment. The artillery regiment, for example, had just one battalion for a long time, and the anti-tank battalion at first had just 12 75-mm heavy anti-tank guns. Only after it had been sent to the Eastern Front did the battalion receive 10 assault guns. If one compares the organization of a panzer grenadier division with that of this panzer grenadier brigade, one notices that neither an independent panzer nor assault gun battalion was envisaged. The unit was thus essentially a motorized grenadier unit.

After the 4th SS Volunteer Panzer Grenadier Brigade "Netherland" arrived at the front around the Oranienbaum pocket in the sector of the 18th Army, in the first days of January 1944 it relieved the battle group of the SS Police Division. In order to deceive the enemy, on 14 January 1944 the commanding general of the III (Germanic) SS Panzer Corps gave an order that the brigade was to be called the SS Volunteer Panzer Grenadier Division "Netherland."

That same day the Red Army opened its winter offensive against the left wing of the 18th Army near Oranienbaum. The next day there was fierce fighting between the attacking Soviets and troops of the 4th SS Volunteer Panzer Grenadier Brigade.[34] After Soviet troops succeeded in encircling elements of the 9th and 10th Luftwaffe Field Divisions, battle groups from the brigade helped break open the pocket, enabling the surrounded German troops to escape.

On 26 January 1944 the brigade received orders to fall back to the so-called Panther Position along the Narva River. After fierce defensive fighting, at the beginning of February 1944 the unit marched through Krikkolo and occupied an interim position on the Luga River. After the units finally moved into the Narva bridgehead there was quiet at the front for a few days. The brigade's first action had ended after about two weeks of heavy fighting and a retreat of almost 100 kilometers.

After several smaller local attacks and counterattacks during February 1944, on 11 March the Soviets again launched heavy attacks on the entire main line of resistance. SS Volunteer Panzer Grenadier Regiment 48 was deployed in the northern sector of the Narva bridgehead and SS Volunteer Panzer Grenadier Regiment 49 in the eastern part of the bridgehead containing the border city. On 15 March 1944 the Wehrmacht communiqué had words of praise:

"The Dutch SS Volunteer Panzer Grenadier Regiment "General Seyffardt" commanded by SS-Obersturmbannführer distinguished itself in the recent fighting in the northern sector of the Eastern Front."

The next day III Battalion, SS Police Artillery Regiment, which had been attached to the brigade for some time, was transferred to the 4th SS Volunteer Panzer Grenadier Brigade as SS Artillery Battalion 54. The brigade thus at least had a battalion of light artillery.

The brigade's losses from 1 January to 31 March 1944:

	Officers	Other Ranks	Total
Killed	25	626	651
Wounded	33	2,244	2,277
Missing	5	473	478
Total	63	3,343	3,406

The brigade's casualties during the first 14 days of April totaled 322 killed, wounded, and missing. In 14 weeks of action the brigade had suffered 3,728 casualties and had to be seen as almost destroyed.

SS Volunteer Panzer Grenadier Regiment 24 "Denmark" of the 11th SS Volunteer Panzer Grenadier Division was in the city of Narva in the southern sector, which had been declared a "fortress." In order to ensure a unified command in action the regiment was finally placed under the commander of the 4th SS Volunteer Panzer Grenadier Brigade "Netherland," whose unit was then renamed Group Wagner.

On 13 May 1944 Group Wagner was itself placed under the commander of the 11th SS Volunteer Panzer Grenadier Division to ensure a more standardized issuing of orders. The front along the Narva had meanwhile grown quiet, and apart from mutual patrolling in the eastern sector of the bridgehead there was no serious fighting. George Duiker recalled:

"When we arrived in Russia from Croatia and detrained I was ordered to take three motorcycle-sidecar combinations to determine where German troops were. First the French motorcycle broke down, then the English one, and I was forced to go on alone on the Zündapp with two soldiers. I had to brace myself in the sidecar with both hands and feet, for my driver drove quite fast despite the snowdrifts. After a long night drive we reached the headquarters of the 11th SS Volunteer Panzer Grenadier Division "Nordland" on the coast. I was immediately asked where the "Nederland" brigade was. All the way back we again heard the sound of fighting. When I returned I was able to tell the Dutch where our troops were supposed to march. We then moved to the front of the Oranienbaum Pocket, in part on foot. Retreating members of the Luftwaffe field divisions (Luftwaffe-Feld-Divisionen)—also called Luftwaffe bad designs (Luftwaffe-Fehl-Konstruktionen)—were incorporated into our units.

Finally we were ordered to withdraw behind the Narva. During the night we moved out and fell back along a coast road. We waited for food, but nothing came forward. That was very bad for the morale and fighting strength of the unit. Then we came to a crossroad outside Narva which we were supposed to hold to allow the retreating units to pass through the swampy area. The Russians pursued hard with motorized units and tanks. Things went full tilt in front of Narva. The Russians attacked without pause. On one occasion an SS-Hauptsturmführer came running toward me wearing just a shirt. He had thrown away his sweater and camouflage tunic. He could barely speak. I had a vehicle take him to the rear. During the day the Russians attacked and shot up a neighboring army battalion. All the officers of the battalion headquarters came running towards us. They left the wounded moaning in the snow.

Then the Russians attacked us with seven T-34 tanks. I was able to disable all of them with my two 75-mm anti-tank guns. Then we pulled back. After a difficult night march through deep snow we reached Hungerburg (Narva-Jõesuu). As the ice on the Narva had been blown up to prevent the Russians from crossing, we had to improvise our way across the river and got completely soaked in the process. In that state we had to push on immediately to the "Nederland" Brigade in the Narva bridgehead!

We then occupied positions on Narva's east bank, and I was given several recruits as replacements. When I asked them which weapons they had trained on, they replied: "In laying cable." I asked again, and it turned out that they were actually signals men. My first task was to instruct them what to do there. After a barrage I saw a soldier huddled up. I rushed over to him, thinking he had been wounded. But no, he was praying and had the rosary in his hands. I soon made him stop that.

At the beginning of February 1944 the Red Army succeeded in establishing a bridgehead southwest of Narva, and I was sent there with my anti-tank gun. On the way my driver put our captured French truck into a skid. It flipped over and I broke my hand. I stopped a truck and, using a tow cable, we got our truck back on its wheels. We towed it out of the snow back onto the road and away we went. The wheels were out of alignment, but it was drivable. During the fighting I was hit by fragments from a hand grenade and was sent to a hospital behind the front. Finally I was flown out in a Ju 52, and in mid-July 1944 I arrived at the 1st SS Anti-Tank Training and Replacement Battalion in Rastenburg. I was there during the assassination attempt on Hitler. My service in the "Nederland" was thus over."

After receiving numerous replacements—again mostly ethnic Germans from Rumania—on 30 June 1944 the brigade reported a strength of:

	Officers	NCOs	Enlisted Men	Total
Actual	220	1,319	5,175	6,714
	3.3%	19.7%	77%	100%
Authorized	325	1,895	6,740	8,960
	3.6%	21.2%	75.2%	100%

Before Army Detachment Narva (*Armee-Abteilung "Narwa"*), and with it the reinforced 4th SS Volunteer Panzer Grenadier Brigade "Netherland," were attacked again the assassination attempt on Hitler took place. *SS-Brigadeführer* Wagner made the following demagogic comments in his order of the day:

"SS men of the Netherland Brigade!

Since being given the task of leading you as Dutchmen, ethnic Germans from Transylvania and Reich Germans, I have repeatedly stressed that the toughest battle was coming, the day of decision.

Now it has come!

Surrounded and totally outnumbered by enemies, a dastardly clique of traitors has tried to treacherously murder the German people, and thus Europe, for a second time. The sign of an act of providence that was given us today in the preservation of the life of our Führer, gives us increased strength to achieve that for which a man has devoted his entire life, a man who is leader of us all.

As a veteran SS man, in this difficult hour which has struck us all, I call out to you the words that has sustained the national-socialist ideal through the most difficult years of struggle and brought it to the apex of its achievements:

Now all the more!

Our lives belong to the Führer –we will remain faithful to him!"

During the course of the major Soviet offensive, on 24 July 1944 the first elements of Group Wagner pulled out of the front to establish a new main line of resistance in the so-called "Tannenberg Position." SS Volunteer Panzer Grenadier Regiment 48, which was acting as rear guard, left the Narva on the night of 25 July. The regiment was trapped by Soviet troops in the forests southwest of Lagna and almost wiped out. A former member recalled:

"The withdrawal of all non-essential trains into the Kothla-Järwe area began on 19 July 1944. Five days later the 3rd Baltic Army attacked Army Detachment Narva. The Ivan achieved a penetration in the Riigi – Hungerburg area. The withdrawal order stated that all units should have crossed Grid Line 61 (Hungerburg – Suur-Soldino) to the west by 18:00 hours on 24 July 1944. SS Panzer Grenadier Regiment 48, plus II Battalion, SS Panzer Grenadier Regiment 49 and 7th Company, SS Panzer Grenadier Regiment 24, remained on the west bank of the city of Narva for a few hours as rear guard. The bridges across the Narva were blown at two in the morning on 25 July 1944. The subsequent retreat by SS Panzer Grenadier Regiment 48 was a complete fiasco. An entire regiment was destroyed because of bad decisions and rashness! While all the other units of the III (Germanic) SS Panzer Corps moved along the railway line into the Tannenberg Position, our regimental commander ordered us to march north to the road. The traffic there was as heavy as at peak vacation time, and it was impossible to make any forward progress. When the Russians noticed us we came under heavy fire. The regiment subsequently pulled back and marched southwest like a herd of sheep, all the while under fire from enemy aircraft. The first signs of disintegration were seen at about 10:00! Everything that slowed our retreat was destroyed. There was total

75-mm Pak 40 anti-tank gun.

Sturmgeschütz IV (Sd.Kfz. 163) armed with the 75-mm StuK 40 L/48.

75-mm Type 18 infantry gun, also referred to as "Gypsy artillery."

confusion and no one knew what to do next. Instead of trying to force a breakthrough with powerful assault teams the men fought in small groups. Some people took their own lives, including SS-Sturmbannführer Breimann (C.O. of II/48). This did nothing to improve morale. Various small groups tried to fight their way out and a few succeeded. Many were killed or captured. Of our 15 men eight were killed in the attempt, and the remaining seven, some of them badly wounded, regained our own positions. That was the end of the "glorious" SS Panzer Grenadier Regiment "General Seyffardt," which was senselessly sacrificed because of an indecisive command."

After the German units had occupied the new position the front ran from the Gulf of Finland about three kilometers to the south to the Narva – Perjatse road. The line was anchored by the so-called Blue Mountains, with Sanatorium Hill in the east, Grenadier Hill to its west, and Hill 69.9. Deployed in the main line of resistance from the 4th SS Volunteer Panzer Grenadier Brigade "Netherland" were SS Volunteer Panzer Grenadier Regiment 49 "de Ruyter" (Dutch No. 2) and SS Pioneer Battalion 54.

On 26 July 1944 the Red Army captured Sanatorium Hill. There was also fierce fighting for possession of Grenadier Hill and Hill 69.9. During these actions I Battalion, SS Waffen Grenadier Regiment 47 (Estonian No. 3) was attached to SS Volunteer Panzer Grenadier Regiment 49.

With the shifting of the Soviet main effort to Army Detachment Narva's southern boundary with the 18th Army, the front before the Tannenberg Position grew quiet. After taking over a battle group[35] for the threatened Dorpat area, *SS-Brigadeführer* Wagner was relieved as commander of the remnants of the 4th SS Volunteer Panzer Grenadier Brigade by *Oberst* Friedrich, the adjutant of Army Detachment Narva.

On 9 September 1944 *SS-Brigadeführer* Wagner was placed back in command of the reinforced remnants of his brigade:

SS Volunteer Panzer Grenadier Regiment 49 "de Ruyter"
SS Volunteer Artillery Regiment 54
SS Pioneer Battalion 54
SS Anti-Tank Battalion 54
SS Armored Reconnaissance Battalion 54
SS Medical Company 54
SS Signals Company 54
SS Economic Company 54
SS Supply Services 54
SS Field Replacement Battalion 54
Battle Group, 5th SS Volunteer Assault Brigade "Wallonien"

IM NAMEN DES FÜHRERS

VERLEIHE ICH
DEM

ϟϟ-Hauptsturmführer

Paul K r a u ß, geb. 16.9.13,

Stu.Gesch.Abt.54

DAS

EISERNE KREUZ
1. KLASSE

Brig.Gef.Std., den 4.8. 1944

ϟϟ-Brigadeführer
und Generalmajor der Waffen-ϟϟ
(DIENSTGRAD UND DIENSTSTELLUNG)

On 4/8/44 SS-Hauptsturmführer Krauß was awarded the Iron Cross, First Class for his actions in the fighting for Grenadier and Sanatorium Hills in Estonia.

I Battalion, SS Waffen Grenadier Regiment 47 (Estonian No. 3)

After developments in the situation in southern Estonia made the encirclement of the German forces in northern Estonia—including the Tannenberg Position—appear likely, on 18 September 1944 Army Group North ordered "Operation Aster," the evacuation of Estonia. The withdrawal from the positions and the march into the Pernau area began that same evening. While the "*Wallonien*" battle group had already been transferred, the attachment of I Battalion, SS Waffen Grenadier Regiment 47 (Estonian No. 3) ended here.

On 21 September 1944 the III SS Panzer Corps ordered the reformation of SS Volunteer Panzer Grenadier Regiment 48. The remnants of the regiment were sent by sea from Pernau to Germany. The regiment was reformed at Training Camp "Hammerstein" in Pomerania, mainly with personnel released by the German Navy. At the end of September 1944 the brigade reported a strength of:

	Officers	NCOs	Enlisted Men	Total
Actual	178	1,170	5,182	6,530
	2.7%	17.9%	79.4%	100%
Authorized	325	1,895	6,740	8,960
	3.6%	21.2%	75.2%	100%

As a result of the transfer of naval personnel, the brigade's personnel strength was scarcely different than it had been on 30 June 1944, despite the fierce and costly fighting. The percentage of Netherlanders in the 4th SS Volunteer Panzer Grenadier Brigade "Netherland" had dropped to only about 15%, or about 1,000 men.

In Latvia the unit was attached to the L Army Corps and initially used as corps reserve. While the other units marched in the direction of Riga, SS Volunteer Artillery Regiment 54 was deployed with the 21st Infantry Division between Burtnek and Wolmar to repel Soviet attacks. A combat report by SS Volunteer Panzer Grenadier Regiment 49 dated 26 September 1944 concerning an action in the Aizani area documents the destruction of at least 13 Soviet tanks with close-range weapons:

"*In the morning hours of 24/9/44 the Petersen battle group occupied the position on the Lemmer. Our reconnaissance carried out during the day had detected no enemy movement. At about 19:30 hours the combat outposts reported the approach of three tanks and 20 trucks loaded with infantry. Two tanks drove along the beach, reached the road south of the Lemme, and continued towards it, and the third tank was destroyed with close-range weapons. The enemy infantry tried to break through our position but were repulsed with heavy losses.*

The enemy then launched a massive armored attack along the coast and toward the road with about 20 tanks. He succeeded in breaking through our main line of resistance. Five more tanks were destroyed with close-range weapons. Several tanks drove to where our own combat vehicles were parked and were engaged there by the reserve and drivers. Four more enemy tanks were destroyed.

During the ensuing pause in the fighting the battalion withdrew under heavy enemy fire. The enemy then tried to overtake the moving column and attack it from behind. SS-Unterscharführer Sporck brought two light infantry guns and, constantly changing positions, destroyed one tank and thus covered the withdrawal.

The two tanks that had broken through in the first attack were knocked out by the 4th Battery, SS Artillery Regiment 54 and the squad leader of the division radio station.

Two enemy tanks suddenly appeared as the light infantry guns were again changing positions. The first rammed the vehicle from behind but was disabled with a grenade. The second rolled along the column. The men from the rammed vehicle ran after the enemy tank and destroyed it from behind with a hand grenade. In order to save the main body of the battalion from further attacks by enemy tanks and prevent the vehicles from being destroyed the commanding officer gave the order to blow the bridge north of Haynasch (Ainizi), even though his adjutant and a large number of men were still on the north side of the river."

After marching about 300 kilometers from Pernau, on 28 September 1944 the troops reached the area south of Doblen and immediately became engaged in heavy defensive fighting. They were reinforced by III Battalion, SS Waffen Grenadier Regiment 34 (Latvian No. 3) and IV Battalion, SS Waffen Artillery Regiment 19 (Latvian No. 2).

At the beginning of October 1944 the Red Army succeeded in trapping Army Group North in Courland. On 12 October the unit, now called simply the SS Volunteer Panzer Grenadier Brigade "Netherland," was ordered into the area south of Libau, approximately 150 km away. The III (Germanic) SS Panzer Corps was to break through to the German main line of resistance in East Prussia. Powerful Soviet forces prevented the operation from being carried out, however. Instead, on 14 October the brigade was attached to the 11th Infantry Division (I Army Corps) to combat enemy forces that had broken into the wooded area north of the Tirs-Purvs marsh. Despite fierce enemy attacks in this area, on 17 October the unit was taken out of action and sent back to the III SS Panzer Corps in the area 10 km south of Prekuln. The SS Volunteer Panzer Grenadier Brigade "Netherland" was subjected to repeated heavy attacks and its lines were broken several times. Consequently, the brigade was forced to withdraw into the area 15 km southwest of Prekuln.

On 1 November 1944 the brigade withdrew to the line from west of Skuodas to Prekuln, as ordered by the 18th Army. It subsequently moved into the so-called Kriemhild Position, which was held until January 1945.

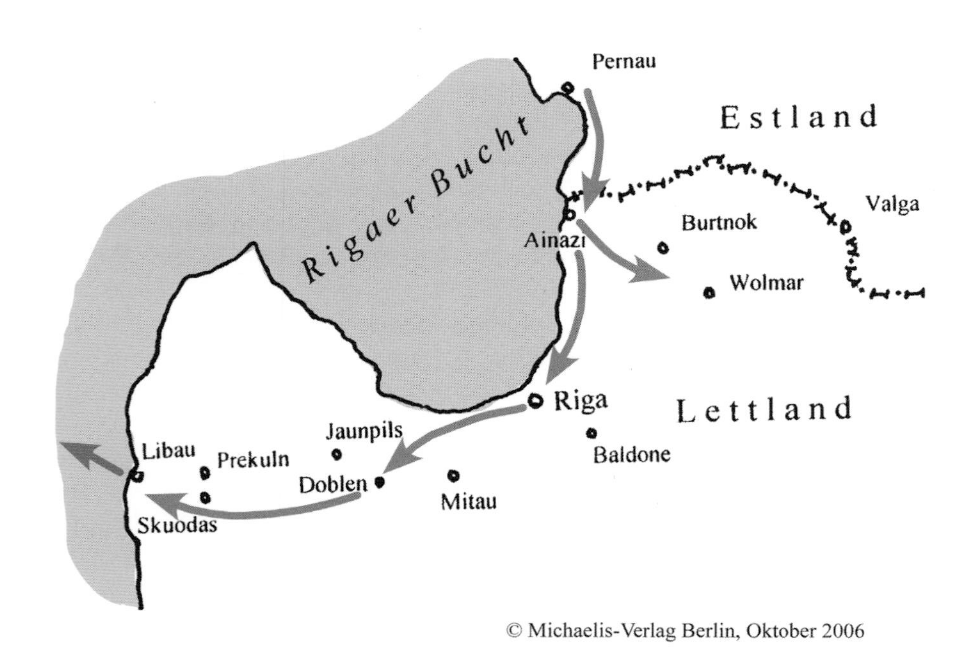

On 16 January 1945 the brigade reported the following strengths:

	Rations strength	Combat Strength
Division Headquarters	248	
SS Vol. Pz.Gren. Rgt. 48	2,166	reorganization
SS Vol. Pz.Gren. Rgt. 49	1,105	797
SS Vol. Art.Rgt. 54	951	567
SS Recon. Comp. 54	67	32
SS Pioneer Btl. 54	253	121
SS Anti-Tank Btl. 54	402	approx. 300
SS Flak Comp. 54	97	67
SS Medical Comp. 54	148	
SS Signals Comp. 54	186	
SS Economic Comp. 54	123	
SS workshop Comp. 54	92	
SS Field Post Office	13	

With a total strength of 5,751 men, the brigade's actual strength was that of two weak infantry battalions or one weak artillery battalion.

After a relatively quiet period, on 21 January 1945 the Red Army launched a massive attack against the entire Libau front. After fierce fighting for possession of Kaleti the SS Volunteer Panzer Grenadier Brigade "Netherland" was pulled out of the main line of resistance and attached to X Army Corps. On 26 January 1945 *SS-Brigadeführer* Wagner documented the remaining combat units:

"*SS Volunteer Panzer Grenadier Regiment 49:*

After its withdrawal, the regiment has no value as a fighting unit. The battalions are down to 10 – 20 men each. Almost all of the heavy weapons are out of action, and there is almost no one available to serve them, as their crews were used in the infantry role after their weapons were disabled.

SS Artillery Regiment 54:

The regiment's personnel strength has continued to drop. It has suffered significant casualties as a result of personnel being used to form scratch infantry units. Although replacement personnel are needed to achieve full operational readiness, the regiment can be considered operational.

SS Pioneer Battalion 54:

The pioneer battalion has no value as a fighting unit. Apart from the adjutant and one company commander it has no pioneer officers or NCOs. Its combat strength is down to 10 – 15 men. The rear-echelon elements are largely intact."

On 28 January 1945, what was left of the units was transported by sea from Libau to Stettin.

The 23rd 4th SS Volunteer Panzer Grenadier Division "Netherland" (Dutch No. 1)
(Die 23. SS-Freiwilligen-Panzergrenadier-Division "Nederland" (niederländische Nr. 1))

After disembarking in Stettin on 2 February 1945, following an order to rename the Waffen-SS brigades divisions the unit was given the title: **23rd 4th SS Volunteer Panzer Grenadier Division "Netherland" (Dutch No. 1)**.

Apart from the fact that the unit was at least 5,000 men below the authorized strength of an Infantry Division 45, with barely a battalion from the Netherlands, its designation as a "Dutch" SS volunteer unit was something of a paradox.

There was no possibility of coming anywhere close to authorized strength, and employment of the unit as one body was out of the question because of the physical separation of the two panzer grenadier regiments. Consequently, two regiment-strength battle groups were created with supporting weapons.

The first unit to see action was the SS Volunteer Panzer Grenadier Regiment 48 "General Seyffardt," which had been in the reformation process at Training Camp "Hammerstein" since October 1944. The unit's I Battalion had been formed from the remnants of the original regiment, while II Battalion was created with troops provided by the SS Panzer Grenadier Instruction Regiment in Neweklau. While I Battalion consisted mainly of ethnic Germans from Transylvania, II Battalion was made up, for the most part, of former members of the naval artillery. There were virtually no Dutch troops left in the regiment, whose title included the suffix "Dutch No. 1."

In mid-January 1945 the Red Army launched a drive towards the west from the Vistula bridgeheads on either side of Warsaw. The regiment was alerted and attached to XVI SS Army Corps. The unit was encircled on 1 February 1945 during the defense of Schneidemühl, and on 14 February the survivors broke through to the German lines. The regiment was then attached to XVIII Mountain Corps and placed under the command of "Group Ax," which consisted of:

- Headquarters, 15th SS Waffen Grenadier Division
- Battle Group Scheibe
  - remnants of Grenadier Regiment 59
  - remnants of SS Volunteer Panzer Grenadier Regiment 48
  - 2 alert battalions
- Battle Group Janums
  - remnants of 15th SS Waffen Grenadier Division
  - 2 alert battalions

The division group had no heavy weapons, and it had to face the Soviet air force, artillery, and armor with only light infantry weapons: carbines, machine-guns, and panzerfausts.

Capable of offering no more than token resistance to the Red Army's advance, the group—now under the command of *SS-Oberführer* Burk—was forced back to Hammerstein on 26 February. There it was attached to Corps Group Tettau, which formed the 3rd Panzer Army's left wing to the Baltic. Initially pulled back to Belgrade, the corps was almost encircled. The attempt to regain contact with the German lines again resulted in chaotic conditions. With no tight control over the units the corps group's commanding general, Generalleutnant von Tettau, saw no other possibility than to give the order to destroy all remaining weapons and vehicles and break through to the west in small groups.

Via Crössin – Zedlin, on 11 March 1945 the remnants of the SS Volunteer Panzer Grenadier Regiment 48 reached the Dievenow bridgehead. Operational elements were temporarily attached to Corps Group Munzel for use in the bridgehead position. On 19 March the men left the bridgehead and moved into the Heinersdorf area (approx. 5 km NW of Schwedt) to be reorganized. Several days later the remnants of the 23rd SS Volunteer Panzer Grenadier Division, also deployed in Pomerania, arrived in the same area.

Following the evacuation of Courland, the bulk of the division with SS Volunteer Panzer Grenadier Regiment 49 "de Ruyter" was immediately attached to the III SS Panzer Corps of the 11th Army. As a replacement for SS Volunteer Panzer Grenadier Regiment 48 "General Seyffardt," on 12 February 1945 SS Panzer Grenadier Regiment "Klotz," formed from the SS Panzer Grenadier School Kienschlag, was attached to the division. SS Reconnaissance Company 54 and SS Flak Company 54 were disbanded effective 5 February 1945. SS Pioneer Battalion 54 received a fully-equipped company from the SS Pioneer School Hradischko. The percentage of Dutch troops in the entire 23rd 4th SS Volunteer Panzer Grenadier Division "Netherland" (strength about 6,000 men) was less than 10%. Most of the rest were ethnic Germans and former members of the navy.

In mid-February 1945 the division Headquarters had under its command:

SS Volunteer Panzer Grenadier Regiment 49 (2 battalions)
SS Panzer Grenadier Regiment "Klotz" (2 battalions)
SS Volunteer Artillery Regiment 23[36]
SS Pioneer Battalion 23
SS Anti-Tank Battalion 23 (30 anti-tank and assault guns)
SS Signals Battalion 23
SS Economic Battalion 23

Hans-Friedrich Naumann

1943 Soldbuch zugleich Personalausweis 1945

Nr. 119/43

für

den Matrosen (Dienstgrad) / 16 (Wehrsoldgruppe)

befördert: ab (Datum)	zum: (neuer Dienstgrad)	(Wehrsoldgruppe)
1.5.43.	Gefreit. (M.A)/S.R	15
9.3.44	M.A. Gefr.	15
20.5.44	[illegible] Gefr.	15
1.12.44	SS-Unterscharführer	14

Naumann - Hans Friedr. (Vor- und Zuname)

Beschriftung und Nummer der Erkennungsmarke: N 1372/43 D

Blutgruppe:

Gasmaskengröße: 2

Marine-Stammrollen-Nr.: N 1978/43 D

Stamm-Marineteil: 2. A. d. N, 2. M.E.A. W'haven

Marine-Ersatzkompanie (M.E.K.) [illegible]

geb. am 13.12.24 in Mülheim / Ruhr (Ort, Kreis, Verw.-Bezirk)

Religion: ev. Stand, Beruf: Schüler

Personalbeschreibung:

Größe: 175 Gestalt: schlank

Gesicht: oval Haar: dkl. blond

Bart: keinen Augen: blau

Besondere Kennzeichen (z.B. Brillenträger): Brillenträger

Schuhzeuglänge: 42 Schuhzeugweite:

Hans-Friedrich Naumann

(Vor- und Zuname, eigenhändige Unterschrift des Inhabers)

Die Richtigkeit der nicht umrandeten Angaben auf Seiten 1 und 2 und der eigenhändigen Unterschrift des Inhabers bescheinigt

30 JAN. 1943

den

4./-16. [illegible] A.

(Feldpost-Nr. der ausfertigenden Dienststelle)

(Eigenhändige Unterschrift, Dienstgrad u. Dienststellung des Vorgesetzten)

Kapitänleutnant M.A. und Kompaniechef

Bescheinigungen

über die Richtigkeit der Zusätze und Berichtigungen auf Seiten 1 und 2

Lfd. Nr.	Art der Änderung	auf Seite	Datum	Dienststelle (Feldpost-Nr.)	Unterschrift	Dienstgrad und Dienststellung
1	Beförderung	1	1.5.43	1/-KAL	[illegible]	Oberleutnant M.A. u. Kompaniechef
2	[illegible]	1	9.3.44	Dienststelle Feldpostnummer 20806	[illegible]	Kapitänleutnant M.A.
3	[illegible]	1	14.6.44	Dienststelle Nr. 31812 A	[illegible]	u. Batt.chef
4	Dienstgrad	1	[illegible]	[illegible]	[illegible]	[illegible]

Inhaber wird überwiesen:

am (Tag der Inmarschsetzung)	an (neue Dienststelle, Feldpost-Nr.)	Gezahlter Reisevorschuß RM [1]		Bescheinigung der abgebenden Dienststelle (Feldpost-Nr.) – Verwaltung durch Stempel und Unterschrift Dienststelle
25.3.44	2. S. St. A. Goningen	—		Feldpostnummer 20806 Leutnant ... 5. Kompanie 2. Schiffsstammabteilung
25.4.44.	16. O. K. 202 Dienststelle	—	—	
1.5.44.	Fp. Nr. 31812 A	—	—	Dienststelle Fp. Nr. 31812 A
22.7.44	an das Heer			
4.12.44	SS-Frw. Pz. Gren. Rgt. 48 "G.S."	3. Kp.		311/44 I.

In das Lazarett mitgegeben:
Geld, geldwerthabende Papiere, Wertgegenstände u. dergl.

Waffen-SS – SS-Freiw. Pz. Gren.-Rgt. 48 „General Seyffardt"

– 1. Dez. 1944

SS-Hauptsturmführer

Lübeck, d. 10.7.45

Hospital 17
Pionierkaserne
Block 3

Entlassungsbefund:

SS-Uscha. Naumann, Hans geb. 13.12.24

Diagnose bei Entl.: Zustand nach Amputation am linken Oberarm.

Beurteilung: v. U. 60. w. U.

W.D.B.: Ja.

Versehrtenstufe: III

Der Stationsarzt:

Der Blockarzt:

Res. Lazarett
Pi. Kaserne Lübeck
Fernspr. App.

(Prof. Heydemann)
Oberstabsarzt

(Dr. Otte.)
Stabsarzt

Matrosengefreite Naumann was taken into SS Volunteer Panzer Grenadier Regiment 48 "General Seyffardt" as an SS-Unterscharführer on 1/12/44. On 24/3/45 he was seriously wounded by an artillery shell and his left arm had to be amputated just before the end of the war.

SS Medical Company 23
SS Workshop Company 23
SS Field Post Office 23

Lacking a panzer battalion and armored vehicles, such as armored troop carriers or motorized artillery, the organization was in no way that of a panzer grenadier division.

On 15 February 1945 *SS-Brigadeführer* Wagner received orders to assemble his units in the Ravenstein – Jacobshagen area in preparation for the imminent Pomerania offensive.[37] The next day the 23rd 4th SS Volunteer Panzer Grenadier Division set out to retake Reetz. The offensive was soon called off, but until then the division was involved in heavy, confused fighting in the Reentz area. The Red Army's superiority in men and equipment prevented the unit from achieving any success.

After several days of quiet, on 1 March 1945 the enemy launched an offensive in Pomerania and was able to smash open a 15-kilometer-wide breach at the seam between X SS and II SS Panzer Corps. Incapable of opposing the powerful attack, the 23rd 4th SS Volunteer Panzer Grenadier Division was forced to withdraw. The units moved through Freienwalde and Daarz, and on 7 March arrived in the area of Gollnow (approx. 25 km NE of Stettin). There it occupied position in the northern area from Lübzin to Hornskrug in the Stettin-Altdamm bridgehead.

In the battle for this German bridgehead, which was supposed to be a jumping off point for an offensive toward Danzig, the German units were almost destroyed. On 20 March 1945 the bridgehead was abandoned. The unit assembled in the area west of Stettin, and one week later the Commander-in-Chief of the 3rd Panzer Army assessed its combat value:

"*For all intents and purposes the division is no more than a reinforced regimental group. At present its infantry fighting strength is almost nil, and it needs to be completely reorganized. It is impossible to predict at this time when refreshment of the unit will be completed, therefore it is only suitable for defensive tasks, Fighting Worth IV.*"

The division suffered a total of 2,014 casualties between 1 and 31 March, equivalent to almost four battalions:

	Officers	Other Ranks	Total
Killed	13	249	262
Wounded	38	932	970
Missing	6	776	782

The high number of missing can be attributed to several factors: there was no cohesion between the hastily-formed units, which consisted to a large degree of soldiers (often ethnic Germans, who had been drafted into the service in the summer of 1944) with little enthusiasm for fighting who quickly fled or surrendered. A former member recalled:

"In mid-February 1945 SS Pioneer Battalion 23 was sent to assembly areas east of Stettin. The enemy was able to gain more ground, and it was obvious that the pioneer battalion would soon be employed in the infantry role. Our 3rd Company

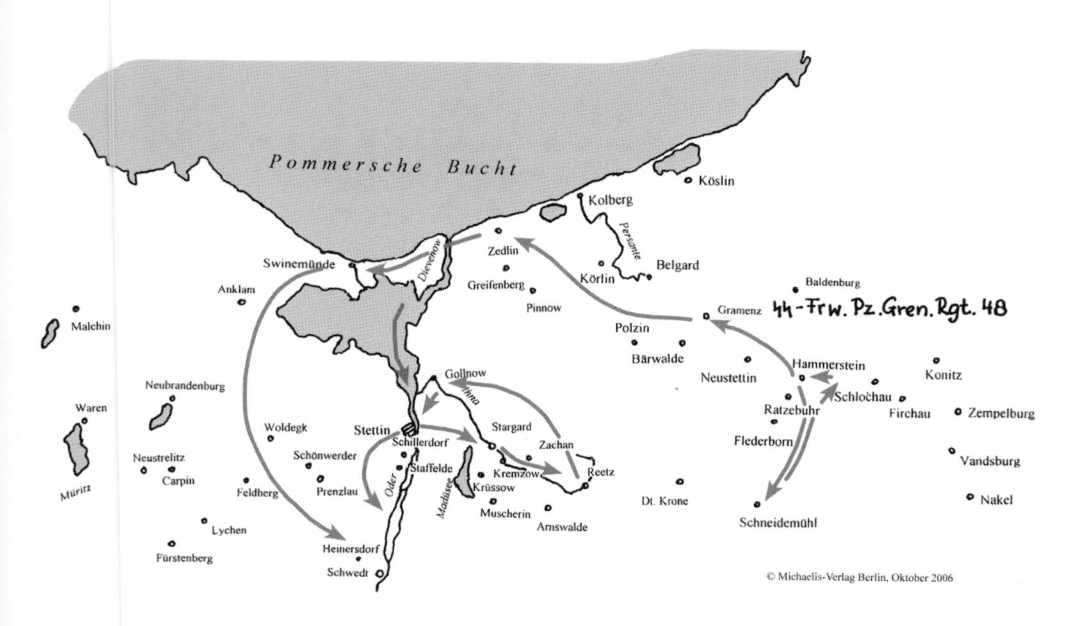

occupied positions on a shallow forward slope south of Ravenstein. Before us was the Ina River. All of this took place with no interference from the enemy. The platoons were familiarized with their sectors without interruption. It was the calm before the storm.

On 1 March 1945 the enemy revealed his offensive intentions, bombarding the battalion's positions with heavy weapons. This lasted all day long. Recovery of the dead and wounded was impossible by day, as the enemy could see every movement. The following night we received the order to pull back. The survivors occupied positions abeam Ravenstein. The enemy had broken through in various places, resulting in considerable confusion. We were finally withdrawn from combat and took up quarters in the houses on the western outskirts of Stettin. After I was wounded and taken to hospital Stabsscharführer Stellmacher brought me many casualty reports, and I can recall having to write letters to about 30 families, informing them that their son or husband had died an honorable death in the battle for Germany. That was equal to about 1/3 of the company on 1 March 1945!

I would also like to say that the "Netherland" Division's well-armed and well-equipped pioneer company never saw a single action in the pioneer role—we were all used strictly as infantry."

The units suffered the following casualties between 1 September 1943 and 28 February 1945:

	Officers	Other Ranks	Total
Killed	55	2,487	2,542
Wounded	141	5,448	5,889
Missing	42	1,609	1,651

The unit had suffered approximately 12,000 casualties in killed, wounded, and missing since its formation in Croatia until 1 April 1945, which meant that theoretically it had been completely destroyed twice. The total number of missing, about 2,400, is noteworthy.

Moved into the area west of Gartz, the remnants of the division were initially attached to the 547th Volks-Grenadier Division. After the SS Panzer Grenadier Regiment "Klotz" had been disbanded due to losses and its remaining personnel divided among the remaining units, these consisted of:

I Battalion, SS Volunteer Panzer Grenadier Regiment 49
SS Volunteer Artillery Regiment 23
SS Pioneer Battalion 23

SS Anti-Tank Battalion 23
  1 Sturmgeschütz III (assault gun)
  1 Sturmhaubitz (assault howitzer)
  4 75-mm anti-tank guns (towed)
SS Signals Battalion 23
SS Economic Battalion 23
SS Medical Company 23
SS Workshop Company 23
SS Field Post Office 23

At the same time, SS Volunteer Panzer Grenadier Regiment 48 was reorganized in the Heinersdorf area. Three battalions were formed using members of the Waffen-SS, Luftwaffe, and Kriegsmarine and sent into the Garz area.

In the final battles that followed the division was again sent into action divided. When the Red Army launched its offensive from the bridgeheads on the Oder on 16 April 1945, the Commander-in-Chief of the 9th Army, *General der Infanterie* Busse, requested the release of the 11th and 23rd SS Volunteer Panzer Grenadier Divisions, which were being refreshed. Intended for the Frankfurt/Oder area, because of transport difficulties both units were initially attached to the equally hard-pressed LVI Panzer Corps in the Wriezen – Seelow area.

I Battalion, SS Volunteer Panzer Grenadier Regiment 49 could not be moved at all, because of lack of fuel, and it occupied positions east of Gramzow (approx. 15 km SE of Prenzlau) in XXXXVI Panzer Corps' sector. Together with II Battalion, SS Volunteer Panzer Grenadier Regiment 69 (Wallonian No. 1), on 26-27 April 1945 the companies were in action northwest of Prenzlau. Under the overall command of the 3rd Panzer Army, the remnants of the battalions marched through Röbel in the direction of Parchim in an effort to reach the demarcation line.

After transport had been organized, the rest of the division with SS Volunteer Panzer Grenadier Regiment 48, which had received replacement personnel, reached the area of the XI SS Army Corps (9th Army) on 18 April 1945. It immediately became involved in fighting in the Marxdorf – Falkenhagen area and withdrew in the direction of Königswusterhausen – Mittenwalde.

On 25 April 1945 the Red Army closed a pocket in which were trapped the remnants of the 23rd 4th SS Volunteer Panzer Grenadier Division "Netherland" (about 4,000 men) and large elements of the 9th Army. With no supplies, air support, or heavy weapons, the days in the so-called Halbe Pocket led to the complete collapse of the units' command and cohesion. Avoiding contact with the enemy, at first the trapped soldiers tried to reach Hermsdorf and then Märkisch-Buchholz using forest roads. From there the columns moved in the direction of Halbe, from where an attempt was to be made to break out to the 12th Army. Countless soldiers and civilians were killed in the breakout attempts that began on 28 April. Those who survived the inferno assembled at the SS reception point in Redekin, northwest of Genthin. There the so-called SS Battle Group Wagner surrendered to the Americans.

Thus ended the story of the 23rd 4th SS Volunteer Panzer Grenadier Division "Netherland" (Dutch No. 1). Out of 8.2 million Netherlanders, in the course of four years a total of about 6,000 men had served in the ranks of the volunteer legion, brigade, and nominal division. After the summer of 1944 the percentage of

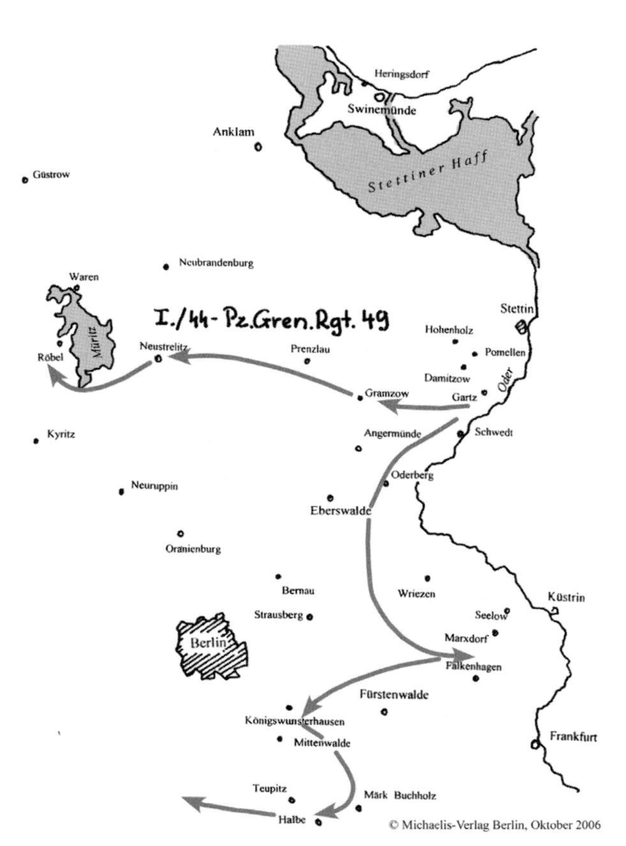

Operations by the Dispersed Division in April-May 1945

Netherlanders was only about 10% of the unit's total strength. The rest were ethnic Germans from Rumania and Reich Germans—many of them former members of the navy. In fact, because of the war situation, neither the brigade nor the division ever attained the full organization of a Waffen-SS panzer grenadier unit. A former member of the division summed it up:

"I signed up with the Volunteer Legion "Netherland" in 1942, probably because, as a 17-year-old, I wanted to become a soldier... and as there was no longer a Dutch army, but recruiting was being done for the war against Bolshevism, I enlisted on the spot. After I joined the replacement unit I began regretting my decision: we Dutch disliked the drill and blind obedience. The harassment by the NCOs and assistant instructors created in us a steadily growing bitterness, and many wanted to get away as quickly as possible. When we were sent to the Legion near Leningrad in 1942, we hoped that things would be better in the core unit. The onset of the miserable autumn, the bad food, and the Soviet winter offensive in early 1943 killed our last motivation. This found its expression in our refusal to volunteer for anything anymore! Instead of being discharged to go home—the original 1 or 2 year enlistments were up for most of us—we were automatically obligated to serve until the war was over. For many that was too much, and we were close to mutiny. In May 1943 we were sent to Grafenwöhr, where a Dutch brigade was supposed to be formed. As the flow of volunteers had dried up because of our letters from the front, ethnic Germans from Rumania had to fill the gaps. There was an outstanding comradeship there with these men and boys, however, they were treated just as we had once been. In many cases friendships developed which have lasted to the present day.

After the Italians went over to the Allies we were sent to Croatia. The conditions there were bad. The broad mass of the rural population, or the population in general, were opposed to their own state, which could not have existed without German troops. And so, while we found ourselves in an allied state, for the most part our presence there was resented. Frequent attacks by partisans made our lives difficult and killed many comrades. And we were used to fighting the partisans, who didn't want us in their country. Paradox. We were happy when at the end of the year we returned to the Eastern Front, which was in fact where we had volunteered to serve.

Morale was not bad—our comradeship with the ethnic Germans was outstanding. We knew that we could trust each other. We arrived at the front, which extended to the west around Leningrad, just in time to prevent the worst—with the result that we suffered terrible casualties in a few days. Not until we were behind the Narva were we able to regroup, reorganize, and rest. The fighting in Estonia in summer 1944 was not easy, but we made the Russians fight for every meter.

At the end of the year in Courland morale again dropped rapidly. The Russians were always stronger than we in winter, and shortages of food and delays in the delivery of weapons and equipment further weakened us. We were frequently urged to desert—in some cases by members of the National Committee for a Free Germany—but for most of us—whether Dutch or ethnic German—their calls fell on deaf ears.

When we were shipped to Pomerania we were happy to be out of miserable Courland—in civilized Germany and closer to our homeland, which had already been occupied by the British! This situation also lowered our spirits. Some of us had lost all contact with our families.

We were constantly amazed at where the Russians got their materiel. If we had 5 – 7 tanks, the other side had at least 50 – 70. Our anti-tank defenses were very weak. Instead of our tanks carrying out attacks, they always had to engage the enemy tanks and—because of lack of fuel—usually came out on the worst end.

After the fighting at Stargard, which was suicidal, we were sent into the area southwest of Stettin. There we received replacements, members of the air force and navy. These men no longer had any motivation. Many had no intention of being killed with the end of the war so near. When the Red Army launched its final attack on 16 April 1945 the "division" was ordered to the southeast. Because of lack of transport and fuel we remained in the rest area (thank God!), and from there marched into American captivity without much fighting. We had survived the war, but we had been on the losing side, and in our Dutch homeland a proper running of the gauntlet initially awaited us. I believe that for many Dutch, the initial period after the war was worse than it had been for most at home during five years of occupation by the German Wehrmacht. For me this included having my teeth knocked out by the Netherlanders who interrogated me! Many of my fellow prisoners endured similar treatment. That was the end of my adventure of fighting against bolshevism! I survived, and through sweat and toil hopefully made possible a good life for me and my family."

Military Postal Numbers

Division Headquarters	57 720
SS Volunteer Panzer Grenadier Regiment 48	40 112
I Battalion	40 899
II Battalion	41 450
III Battalion	42 278
SS Volunteer Panzer Grenadier Regiment 49	59 176
I Battalion	56 387

II Battalion		58 547
III Battalion		57 711
SS Volunteer Artillery Regiment 23	56 420	
I Battalion	04 528	
II Battalion	57 362	
III Battalion		59 048
SS Armored Reconnaissance Battalion 23	56 084	
SS Pioneer Battalion 23	58 308	
SS Medical Company 23	59 719	
SS Signals Company 23	58 949	
SS Economic Company 23	59 884	
SS Division Supply Services 23	57 881	
SS Field Replacement Battalion 23	40 004	

Commanding Officers

07/42	SS-Obersturmbannführer Theuermann
07/42 – 10/43	SS-Obersturmbannführer Fitzhum
11/43 – 05/45	SS-Brigadeführer Wagner

Wearers of the Knight's Cross of the Iron Cross

20/02/43	SS-Sturmmann Mooyman
12/02/44	SS-Hauptsturmführer Dr. Rühle von Lilienstern
21/04/44	SS-Obersturmbannführer Joerchel
04/06/44	SS-Hauptsturmführer Frühauf
04/06/44	SS-Obersturmführer Scholz (24/09/44 Oak Leaves)
19/08/44	SS-Obersturmbannführer Collani
23/08/44	SS-Rottenführer Bruins
23/08/44	SS-Hauptsturmführer Ertel
23/08/44	SS-Sturmbannführer Schlüter
27/08/44	SS-Hauptsturmführer Wanhöfer
02/09/44	SS-Hauptsturmführer Meyer
16/11/44	SS-Rottenführer Strapatin
26/11/44	SS-Standartenoberjunker Schluifelder
11/12/44	SS-Hauptsturmführer Petersen
11/12/44	SS-Untersturmführer Rieth
18/12/44	SS-Kanonier Jenschke
29/12/44	*SS-Brigadeführer* Wagner (Oak Leaves)
05/03/45	SS-Obersturmführer Behler
17/03/45	SS-Obersturmführer Behler
05/04/45	SS-Sturmbannführer und Major der Schupo. Hofer

Appendices

Tables of Organization

SS Assault Brigade (New Type)
SS-Sturmbrigade (neuer Art)
Authorized Strength approx. 2,000 men

Brigade Headquarters
Motorized Grenadier Battalion
    1st – 3rd Rifle Companies
    4th (Heavy) Company
Heavy Motorized Battalion
    5th (Infantry Gun) Company
        2 light and 1 heavy platoons
    6th (Anti-Tank) Company
        1 platoon 50-mm anti-tank guns
        2 platoons 75-mm anti-tank guns
    7th (Assault Gun) Battery
        10 assault guns
    8th (20-mm flak) Battery
    9th (88-mm flak) Battery

Panzer Grenadier Division 43
Authorized Strength approx. 14,000 men

Division Headquarters
2 Panzer Grenadier Regiments
    (one armored battalion)
    I Battalion
    II Battalion
    III Battalion
Motorized Artillery Regiment
    I Battalion (towed)
        1st – 3rd Batteries (105-mm field howitzers)
    II Battalion (towed)
        4th – 6th Batteries (150-mm field howitzers)

III Battalion (towed)
7th – 9th Batteries (105-mm field howitzers)
IV Battalion (towed)
10th – 12th Batteries (150-mm heavy field howitzers)
Armored Reconnaissance Battalion
1 armored car company
2 Volkswagen companies
1 heavy company
Panzer Battalion
4 companies (assault gun)
Anti-Tank Battalion
1 company (14 75-mm or 7.62-mm anti-tank guns) (s.p.)
1 company (12 75-mm anti-tank guns)
Flak Battalion (Towed)
1 battery (12 20-mm anti-aircraft guns)
1 battery (9 37-mm anti-tank guns)
1 battery (4 88-mm anti-aircraft guns)
Pioneer Battalion
3 companies
Motorized Signals Battalion
1 company (field telephone)
1 company (radio)
Motorized Supply Services
3 workshop companies
1 supply company
9 small supply transport columns
3 large fuel transport columns
Administrative Services
1 rations office
1 bakery company
1 butcher company
Motorized Medical Battalion
2 medical companies
3 ambulance platoons

Comparison of SS and Army Ranks

SS-Dienstgrad	**Army**
SS-Grenadier	Grenadier
SS-Sturmmann	Gefreiter
SS-Rottenführer	Obergefreiter
SS-Unterscharführer	Unteroffizier
SS-Scharführer	Unterfeldwebel
SS-Oberscharführer	Feldwebel
SS-Hauptscharführer	Oberfeldwebel
SS-Untersturmführer	Leutnant
SS-Obersturmführer	Oberleutnant
SS-Hauptsturmführer	Hauptmann
SS-Sturmbannführer	Major
SS-Obersturmbannführer	Oberstleutnant
SS-Standartenführer	Oberst
SS-Oberführer	no comparable rank
SS-Brigadeführer	Generalmajor
SS-Gruppenführer	Generalleutnant
SS-Obergruppenführer	General
SS-Oberstgruppenführer	Generaloberst

Bibliography

Federal Archive Berlin:	various personal files
Federal Military Archive Freiburg:	various file collections
Federal Archive Coblenz:	Reichsführer-SS file collection
Eyewitnesses:	various interviews and manuscripts

Notes

1 This was also introduced by the 7th SS Volunteer Mountain Division "Prinz Eugen."
2 See also: Michaelis, Rolf: Der Einsatz der Ordnungspolizei 1939-1945, Berlin 2008.
3 In his duty calendar Himmler noted that the reasons for incorporating the unit into the Waffen-SS as difficulties with the Order Police headquarters and Daluege's constant interference with the command of the division. When the division became part of the Waffen-SS its members were assigned equivalent Waffen-SS ranks. A Gefreite became an SS-Sturmmann, for example.
4 In December 1942, II Battalion, SS Police Infantry Regiment 3 was moved into the Boguchar area (NW of Stalingrad) as part of SS Battle Group Fegelein. In February 1943, with only about 200 of its original 1,000 men left, it was transferred from Valuiki back to Debica.

I Battalion, SS Police Infantry Regiment 1 was also ordered to the hard-pressed Army Group B in December 1942. The MPP were assigned to various battle groups southwest of Stalingrad. Not until 19 February 1943 was the battalion united in SS Battle Group Schuldt. It then consisted of:

SS Guard Battalion "Berlin"
I Battalion, SS Panzer Grenadier Regiment "Der Führer"
I Battalion, SS Police Infantry Regiment 1

SS Battle Group Schuldt was disbanded after heavy fighting at Millerovo and Utkino. I Battalion, SS Police Infantry Regiment 1 returned to SS Training Camp "Heidelager" (Debica) with 84 of its original 527 men.

5 The core of the unit was formed by the SS Assault Gun Battery "Wallonien."
6 In the future the division was described as having two panzer grenadier regiments. SS Police Panzer Grenadier Regiment 3, which was still deployed in Russia, was later divided among the first two regiments.
7 The three companies were under the command of Waffen-Hauptsturmführer Obitz. After he was wounded, Waffen-Hauptsturmführer assumed command of what was left. From Hela the French troops proceeded to Copenhagen, and on 10 April 1945 arrived in Neustrelitz, where the 33rd SS Waffen Grenadier Division "Charlemagne" (French No. 1) was supposed to be reformed.
8 On 16/4/45 the SS Panzer Grenadier Regiment was reorganized and merged with SS Field Training Regiment 103. From then on it was sometimes also called SS Panzer Grenadier Regiment 103.
9 As Denmark was not occupied by German troops after hostilities, this de facto legion was designated a free corps (Freikorps).
10 On 15 September 1943 there were 20 Norwegian officers, 50 NCOs, and 464 enlisted men, plus 33 Danish officers, 162 NCOs, and 1,191 enlisted men—half of them ethnic Germans from North Schleswig—in the division.

11 However, the III Battalion was an armored unit equipped with armored troop carriers.
12 It is difficult to understand the reasons that led Himmler to bestow the name of the Grand Master of the German Order upon the panzer battalion, a unit which had no combat experience. There is nothing to indicate that his decision was connected to the Nordic ideal or the settlement of ethnic Germans in Rumania.
13 They did not, however, return to the command of the 11th SS Volunteer Panzer Grenadier Division "Nordland"; instead, in November 1944 they were attached to the badly-decimated IV SS Panzer Corps.
14 These were Hill 69.9, the so-called "Grenadier Hill," and Sanatorium Hill.
15 See Chapter: Germanic Volunteers – Britons
16 This consisted of three weak (panzer) grenadier battalions and the remnants of other division elements.
17 He was a stickler for regulations, even insisting that the men in the trenches shave every day!
18 The Motorized Grenadier Battalion became II Battalion, SS Panzer Grenadier Regiment 35; the 7th (Assault Gun) Battery became 3rd Company, SS Assault Gun Battalion 16; the 8th (20-mm) Battery became 1st Company, SS Flak Battalion 16; and the 9th (88-mm) Battery became 3rd Company, SS Flak Battalion 16.
19 SS Panzer Battalion 16 was also supposed to be equipped with assault guns.
20 This ultimately formed the basis of a new Escort Battalion "*Reichsführer-SS*."
21 The name "*Rest and Reequipment Group 13th SS Division*" was selected in order to conceal the movement of new troops to Hungary from enemy intelligence.
22 A battalion with more than 400 men was designated "strong"; "medium strong" was 300 to 400 men, "average" 200 to 300 men; "weak" 100 to 200" men, "exhausted" less than 100 men.
23 According to the table of organization, an assault gun battalion was not envisaged as being part of a panzer grenadier battalion. The battalion was formed as a temporary replacement for a usable panzer battalion. When the assault gun battalion was disbanded most of its personnel and vehicles were assigned to the panzer battalion.
24 Götz von Berlichingen—also known as the "Knight with the Iron Hand"—lived from 1480 to 1562, and during the Peasant's War of 1525 he led the rebels in the Oden Forest.
25 Also see the chapter SS Volunteer Panzer Grenadier Division "Nordland."
26 The battalion did not join the division until February 1945.
27 Horst Wessel wrote the SA song "*Die fahnen hoch, die Reihen fest geschlossen...*"
28 The division's new "Volunteer " (*Freiwilligen*) title was added after the bulk of the division was formed with ethnic Germans from Hungary whose fitness grade was not up to that of the SS, but did meet the army standard.
29 The battalion remained part of the 4th SS Police Panzer Grenadier Division and moved with it to Pomerania.
30 On 30 November 1943 the name "General Seyffardt" was transferred from the 1st Company to the first regiment, which on 22 January 1944 also received the suffix (Dutch No. 1).